Chemistry and Life in the Laboratory: Experiments

SIXTH EDITION

Victor L. Heasley
Point Loma Nazarene University

Val J. Christensen
Point Loma Nazarene University

Gene E. Heasley
Southern Nazarene University

Boston   Columbus   Indianapolis   New York   San Francisco   Upper Saddle River
Amsterdam   Cape Town   Dubai   London   Madrid   Milan   Munich   Paris   Montréal   Toronto
Delhi   Mexico City   São Paulo   Sydney   Hong Kong   Seoul   Singapore   Taipei   Tokyo

Editor in Chief: Adam Jaworski

Executive Editor: Jeanne Zalesky

Senior Marketing Manager: Jonathan Cottrell

Assistant Editor: Lisa R. Pierce

Managing Editor, Chemistry and Geosciences: Gina M. Cheselka

Production Project Manager, Science: Maureen Pancza

Cover Designer: Seventeenth Street Studios

Operations Specialist: Jeffrey Sargen

Cover Image: Shutterstock

Credits and acknowledgments borrowed from other sources and reproduced, with permission, in this textbook appear on the appropriate page within the text.

1 2 3 4 5 6 7 8 9 10—VHC— 15 14 13 12 11

PEARSON

www.pearsonhighered.com   ISBN-10: 0-321-75160-4; ISBN-13: 978-0-321-75160-7

Contents

Correlation Page

Chapters in "Fundamentals of General, Organic, and Biological Chemistry, 7e"
by McMurry, Ballantine, Hoeger, and Peterson

Experiments in "Chemistry and Life in the Laboratory"
by Heasley, Christensen, and Heasley

	Principal Experiments	Alternative Experiments
1. Matter and Measurement	1, 5	
2. Atoms and the Periodic Table	6	7, 3
3. Ionic Compounds	6, 7	16
4. Molecular Compounds	3, 7	8, 18
5. Classification and Balancing of Chemical Reactions	6, 7, 3	17
6. Chemical Reactions: Mole and Mass Relationships	11, 5	14
7. Chemical Reactions: Energy, Rates, and Equilibrium	9, 10	4
8. Gases, Liquids, and Solids	2, 4	18, 8
9. Solutions	16, 8	4, 9
10. Acids and Bases	12, 13, 14, 15	
11. Nuclear Chemistry		
12. Introduction to Organic Chemistry: Alkanes	19, 18	
13. Alkenes, Alkynes, and Aromatic Compounds	19, 18	
14. Some Compounds with Oxygen, Sulfur, or a Halogen	20	
15. Amines	23	
16. Aldehydes and Ketones	21	
17. Carboxylic Acids and their Derivatives	22, 23, 25	
18. Amino Acids and Proteins	30, 31	33, 34
19. Enzymes and Vitamins	31, 24	
20. The Generation of Biochemical Energy		
21. Carbohydrates	26, 27, 32	33
22. Carbohydrate Metabolism		
23. Lipids	28, 29	32, 33
24. Lipid Metabolism		
25. Nucleic Acids and Protein Synthesis		
26. Genomics		
27. Protein and Amino Acid Metabolism		
28. Chemical Messengers: Hormones, Neurotransmitters, and Drugs		
29. Body Fluids		16, 24

Preface

In this edition, as in the previous five, our aim has been to provide interesting, provocative experiments that are reliable, unambiguous, and workable. The manual is well suited for use with any introductory chemistry text serving the needs of future health science professionals and general education students. It has been designed to be particularly complementary to the text written by John McMurry, David S. Ballantine, Carl A. Hoeger, and Virginia E. Peterson, *Fundamentals of General, Organic, and Biological Chemistry, 7e* published by Pearson.

This manual will meet the needs of the following students and majors: nursing, paramedical, agriculture, family and consumer sciences, physical education, and general education. The chemistry experiments presented here require students to solve problems that will be directly related to their future occupations. The skills, understanding, and habits of mind that are developed in this manual by making observations, recording data, and drawing conclusions from the data will serve them well as they pursue their chosen professions. We all know that students get excited about a laboratory course when they can relate the experiment to their everyday lives, interests, and careers. For example, in the polymer experiment, students can hardly believe that they are actually making nylon! The experiments are divided among the fields of general, organic, and biological chemistry, a distribution that complements most leading texts.

Numerous changes have been made in this edition, based on our own experience with the manual in the laboratory, observations from the users of the previous editions, and suggestions from reviewers. New to this edition are many short articles, entitled **Environment, Culture, and Chemistry,** directing the student's attention to important topics in the environment and culture that relate to the experiment. Each experiment has been given numbers and letters, correlating the experimental procedures and the questions, making it easier for the student to manage the experiment/question process. The Pre-Laboratory Questions have been collected on a separate, perforated page, which can be removed and handed in at the beginning of the laboratory. Extensive emphasis has been placed on waste disposal. Following each experimental procedure, the student is reminded to properly dispose of the waste, usually upon the direction of the laboratory instructor. The Laboratory Techniques section has been completely revised so that it will make a greater, direct contribution to the student's experimental work. Discussion sections have been modernized with contemporary issues and language. Errors that have come to our attention in the narrative and experiments have been removed or revised.

Across the years, we have become convinced that laboratory exercises should provide a carefully selected variety of representative experiences but need not illustrate every important chemical truth. Specific manipulations and observations may be soon forgotten, but the experimental process ought to make a lasting impression. We believe that experiments should draw heavily from familiar, easily understood situations. They should stress concepts and ideas that will expand the students' understanding of and appreciation for the world around them. Finally, experiments ought to contribute toward enlightening our future decision makers and opinion molders as well as developing necessary skills for the science-related professions.

We have designed the experiments in this manual with these principles in mind. They address a number of specific areas.

1. **Learning basic concepts of matter and energy.** Some novel experiments give insight into the particulate nature of matter, what actually constitutes a chemical reaction, the making and

breaking of chemical bonds, energy aspects of chemical reactions, quantitative relationships in chemical reactions, and some differences between ionic and covalent compounds.

2. **Introducing the chemistry of life.** After students have been introduced to some basic organic reactions, a number of pertinent topics in biochemistry are considered, including the reactions of carbohydrate metabolism, the structure of lipids, the reactions involved in protein digestion, and how enzymes catalyze reactions.

3. **Appreciating the relationship between chemistry and medicine.** Use is made of such medically related examples as iodine chemistry and goiters, enzyme structure and heavy metal poisoning, aspirin structure and fever reduction, fatty acid unsaturation and cholesterol levels, and genetics/enzyme relationships and intolerance to milk.

4. **Understanding familiar materials.** Common materials such as paraffin, turpentine, stomach antacids, bleaches, bouillon cubes, aspirin, nylon, milk, and the peanut are used as a basis for studying the reactions of hydrocarbons, titration, acids and bases, oxidation and reduction, the chemistry of proteins and amino acids, and the synthesis of important material.

5. **Confronting concerns of society.** A number of experiments raise questions and help students develop a basis for enlightened judgment about topics of current interest such as truth in advertising, pollution of the environment, fluoridation of water, and the nutritional value of food.

6. **Developing skills in quantitative measurement.** The experiments dealing with metric measurement, energy calculations, compound formulas, and the yield in the synthesis of aspirin emphasize important aspects of measurement skills and raise provocative questions concerning some of the difficulties encountered in obtaining accurate measurements.

7. **Learning to draw conclusions from experiments and to evaluate the validity of data.** We think the chemistry laboratory is a good place to consider important concepts whose significance transcends the study of chemistry (or science, for that matter) and that are valuable to all professionals and managers. Some experiments are specially designed to teach self-reliance in making measurements and to show that valid conclusions can be drawn from experimental data. On the other hand, we point out that there are limitations to reliability. Since many practical operations and decisions involve cooperative effort, we have included an experiment that illustrates some of the strengths and weaknesses of the team approach.

8. **Examining aspects of advanced chemistry.** The experiment on optical activity introduces students to the exciting world of the professional research chemist.

To help students acquire a suitable foundation, we have given each experiment a list of objectives and several pre-laboratory questions that encourage students to read the discussion thoroughly and to understand the purpose, principles, and procedure of each experiment. Students should thus be able to conduct experiments and make observations with maximum understanding. These aids can greatly reduce the time required for pre-laboratory instruction.

Since experimental observations are of little value unless they are interpreted and their most important meanings assessed, the questions asked in each experiment lead students, a single concept at a time, to probe the meanings and implications of the experiment. At the same time, we avoid the simplistic, lock-step "cookbook" approach. Finally, the Questions and Problems for each experiment extend the implications of the experiment to some practical applications.

Instructor's manual. A thoroughly revised *Instructor's Manual,* available through the publisher, provides a list of chemicals and supplies required for each experiment; suggestions and precautions for optimum class results; and answers to all Pre-Laboratory Questions, selected conclusions in the Observations and Results section, and all Questions and Problems.

Victor L. Heasley

Val J. Christensen

Gene E. Heasley

New to This Edition

- Short articles, entitled **Environment, Culture, and Chemistry** have been added to many experiments, directing the student's attention to important environmental and cultural topics that relate and enhance the experiment. This new feature encourages students to think about chemistry outside of the classroom setting.
- Each experiment has been given numbers and letters, correlating the experimental procedures and the questions, making it easier for the student to manage the experiment/question process.
- The Pre-Laboratory Questions now appear on a separate perforated page, which can easily be removed, handed in, or used in a discussion at the beginning of the laboratory.
- Several new and modified Pre-Laboratory Questions, Observation and Results sheet items, and Questions and Problems.
- Bolded statements that place extensive emphasis on lab safety and waste disposal. Following each experimental procedure, the student is instructed to properly dispose of the waste.
- The Laboratory Techniques Section has been completely revised so that it will make a greater, direct contribution to the student's experimental work.
- Discussion sections have been modernized with contemporary issues and language for greater relevance to current topics.

To the Student

After years of working with beginning chemistry students, we are convinced that good laboratory experiences can make a significant contribution to your professional work and expand the meaning and enjoyment of your everyday life. Many of these experiences are peculiar to this course; they may not be confronted anywhere else in your college career.

In this laboratory course, you will:

1. Apply theoretical ideas to practical situations—a critical aspect of most careers.

2. Acquire skills in performing certain operations, often detailed and complex, to a successful conclusion, and thereby develop confidence in your own abilities and the reliability of scientific experimentation.

3. Learn to make sound conclusions based on your own observations of experimental data.

4. Confront the dynamics of arriving at a decision as a member of a cooperative group.

5. Develop a basis and a habit of thought for making judgments on issues affecting our environment and well-being.

Let us give you some suggestions on how to prepare for each laboratory experience. Each experiment consists of five or six sections: **The Objectives; Discussion; Environment, Culture, and Chemistry** (where applicable); **Pre-Laboratory Questions; Experiment; Questions and Problems; Observations and Results.** We recommend that you approach each experiment by first reading the **Objectives**, and then, with these in mind, read the **Discussion** and **Experiment** sections several times. The **Discussion** is designed to give you background for the particular experiment. Prior to doing the experiment, your instructor will probably require you to answer the **Pre-Laboratory Questions.** Answering these questions will give you a clearer insight into the background and purpose of the experiment.

Read over the questions in the **Observations and Results** section so that you are aware of what will be expected of you for each part of the experiment. *This is essential for a successful lab experience.* Never do an experimental operation without first being familiar with what you are to observe or record. Always answer the questions that pertain to a particular procedure immediately, while the relevant thoughts are still fresh in your mind. Write on the answer sheets as you go along. Never do the entire experiment first and then answer the questions. *Let the questions guide your thinking as you progress through the experiment.*

The **Environment, Culture, and Chemistry** sections are provided to stimulate your thought and increase your understanding of ways that aspects of the experiment apply to our culture. We hope that you find them interesting and helpful.

Take time to read thoroughly the **Laboratory Techniques** section; we recommend that you examine this section several times throughout the course. Your instructor will probably occasionally direct your attention to relevant portions of this section.

Let us discuss with you some philosophical, and yet practical, concepts that may be of benefit as you do the experiments and as you approach life in a complex, technical world where many voices shout conflicting positions.

ATTITUDE

You certainly should not approach this course with a careless, blasé attitude nor should you be paralyzed by fear. One should be excited about exploring the unknown without being careless or presumptuous. A little fear can be a good thing. You need to be alert and have genuine respect for possible chemical and other laboratory hazards. Learning to do things correctly, with understanding, will give you confidence. Scientists in chemical laboratories have developed a set of safety precautions, and strict adherence to these precautions has made the chemistry laboratory one of the safest places on earth to work.

RISKS

You can be assured that in designing the experiments in this laboratory manual, we have given careful attention to maximizing safety and reducing risk, including using minimal amounts of chemicals, most of which have low toxicity. Frequently throughout the book, safety warnings will be given—they must be heeded. We are not saying that all risk has been eliminated—zero risk does not exist in any aspect of life—but we can assure you that the risk level in these experiments is very low.

Some chemicals are very toxic to breathe or ingest, corrosive to the skin, or sensitive to detonation procedures. Others are quite benign, especially in the manner and amounts that are normally encountered in the laboratory. You are already aware of this great disparity in the toxicity of materials, since all that you touch and eat is chemical in nature, and yet most are not toxic. Since you do not know the hazard presented by new and strange chemicals, you must assume that they are dangerous, so do not inhale, ingest, or expose yourself to them unless you are given clear instructions to do so.

QUANTITIES AND CONCENTRATIONS

With experience, you will develop a sense of judgment about hazard and risk that takes into account quantity, concentration, and time. When we say a chemical is toxic, we generally mean that it can cause stress or damage to the body. However, the effect can be greatly influenced by the amount, concentration, and in some cases, time of exposure. Also, some toxic chemicals, such as hydrogen sulfide (rotten egg gas), have effective warning odors, so they can be avoided before they become dangerous. Other hazards, like radiation and the odorless gas carbon monoxide, give no warning.

To illustrate the differences made by amounts or dosages, consider the common gases chlorine and ammonia. Almost everyone has sniffed small amounts of chlorine in a treated swimming pool, but chlorine in large concentrations, particularly over time, is deadly. In fact, chlorine was used as a poison gas against U.S. troops in World War I. Sometimes people who have lost consciousness are given smelling salts (a whiff of ammonia gas) to revive them, yet if you were forced to breathe concentrated ammonia vapors, they would quickly become lethal. Actually, many relatively safe materials can do damage when taken in excess, and others that are toxic are far less hazardous when present in small quantities, at low concentrations, or for a brief time (e.g., X-rays).

A word of warning about toxicity is in order: another reason an occasional whiff of chlorine or ammonia is not dangerous is that the body gradually eliminates these and many other toxic materials. But some chemicals, notably the heavy metals like mercury, are cumulative poisons and the body does not discard them over time. Cumulative poisons should be handled with extreme caution, and we should show extra care in disposing of them in the environment.

WASTE DISPOSAL

For the protection of our environment, we should err on the side of caution when disposing of laboratory wastes. Disposal of chemical waste is of extreme importance in this course. Therefore, we have placed a thorough statement on disposal in the **Laboratory Techniques** section. Furthermore, we have indicated in **bold** print on many occasions throughout each experiment when to dispose of waste; the appropriate containers will be identified by your instructor.

Laboratory Techniques

It is essential that you develop appropriate and professional laboratory practices and techniques so that your experiments will give the anticipated results and that you and your neighbors will be safe. Some important procedures are listed below. You will repeat most of them many times. Refer to them often. Safety considerations are also discussed. Because of their importance, please review these suggestions several times throughout the course as you carry out the various experiments. Professional chemists take safety seriously. Consequently, chemistry research laboratories are among the most safe workplaces.

1. General Procedures and Precautions

a. Only *authorized* experiments may be conducted. Although exploring the unknown may look exciting, it is potentially dangerous at your present stage of learning.

b. Always wear protective goggles or eyeglasses in the laboratory. If a corrosive chemical should get into your eye, flush it immediately with water for at least 15 minutes at an eyewash fountain and immediately notify your instructor.

c. Never drink beverages or eat food materials in the laboratory. Your hands may have traces of chemicals on them and they could be transferred to your mouth.

d. If a corrosive chemical touches your skin, wash it immediately with plenty of water and notify your instructor.

e. Never taste a chemical unless directed to do so by your instructor. Assume all chemicals may be poisonous.

f. Report any injuries, even minor ones, to your instructor.

g. Always maintain a clean work area as you conduct experiments. Clean equipment as you proceed through the experiment. A messy work area can easily lead to confusing results.

h. Wash glassware with water, and a brush and clearer if required. Your instructor will provide directions for where to put clean glassware at the end of the experiment.

2. Handling Reagent Chemicals

a. Read the label on the bottle twice before using.

b. Never return an unused chemical to a reagent bottle so that everyone can be confident of its purity. Try not to withdraw excessive amounts of chemicals.

c. Leave reagent bottles at their designated area. Transport reagents to your desk in beakers, flasks, or test tubes.

d. Most reagent bottles will be equipped with their own dropping pipet or spatula. Never use your own equipment because of contamination problems.

e. When weighing chemicals, be careful to avoid spills. If you do spill a chemical or weigh an excess, consult your instructor for appropriate clean-up and disposal procedures.

3. Smelling Chemicals by Wafting the Vapor

Frequently, you will be instructed to smell a particular chemical and describe its odor. Never smell directly at the mouth of the bottle, flask or test tube, but waft the vapor from the source towards your nose with your hand cupped to move the vapor. During this procedure you should not place your nose closer than four inches to the mouth of the container. (LT-1)

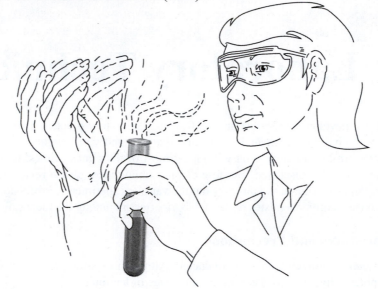

▲ **LT-1:** Wafting a Vapor

4. Heating a Solution in a Test Tube with a Bunsen Burner

When using a test tube holder, hold the test tube at an angle, slowly passing the solution back and forth at the tip of the flame. Frequently pass the flame up the sides of the test tube with shaking to mix the solution. (LT-2). While heating, never point a test tube toward another person. (LT-1)

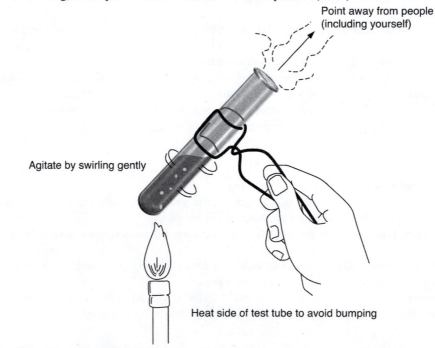

Point away from people
(including yourself)

Agitate by swirling gently

Heat side of test tube to avoid bumping

▲ **LT-2:** Heating a Liquid in a Test Tube

5. Waste Disposal

Proper disposal of chemical waste is an extremely serious issue in this course. Therefore, disposal statements are inserted in bold following most chemical reactions. Normally, these statements will indicate when to dispose of waste in the appropriate containers (identified by your instructor), in a few situations the directions will be specific. Generally the following rules apply: dilute solutions of acids/bases and salts of sodium/potassium chloride/sulfate can be washed down the drain with running water; all other inorganic materials (solids and solutions) must be placed in designated containers; all hydrocarbons (solids and liquids), containing no acids or salts/water, can be mixed in a designated container; hydrocarbons mixed with one or more of the following may require separate containers: water, water/salts, acids, acids/salts; different containers may be needed for individual aqueous salt solutions, depending on their reactivity with each other.

6. When To Use Tap Water and Distilled Water

Tap water may be used in an experiment unless distilled water is specified.

7. Use of pH, Litmus, and Other Test Papers

Tests with pH, litmus, and other test papers should be conducted by transferring a drop of solution on a stirring rod to the paper. Do not put the test paper into the solution since the color can be leached into the solution and the test will not be accurate.

8. How to Properly Stir a Solution

A stirring rod is required to properly mix solids/liquids and liquids/liquids in test tubes, flasks and beakers; shaking or tapping give insufficient mixing in most situations.

9. Drying Glassware

It is frequently necessary to dry glassware during an experiment. Paper towels work well. In the case of test tubes, roll a piece of paper into a pencil-shape, and then twist it into the test tube until it reaches the bottom and removes the water.

10. Measuring the Volume of Solutions

All volumes of solutions should be measured with a graduated cylinder unless you are instructed differently in the laboratory manual or by your instructor. **Beakers**, even those showing volume levels, should not be used in laboratory measurement since they are not sufficiently accurate. Pipets, burets, and volumetric flasks are used for highly accurate measurements. Your instructor will demonstrate the proper use of these pieces of glassware.

11. Pictures of Laboratory Glassware and Equipment

Picture of common laboratory glassware and equipment that you may use throughout the course are shown in LT-3.

Top loading balance

Erlenmeyer flasks

250 mL

100 mL

Büchner funnel

Evaporating dish

Watch glass

Beakers

250 mL

250 mL

150 mL

50 mL

Test tubes

Graduated cylinders

Buret

Volumetric flask

Thermometer

Crucible with cover

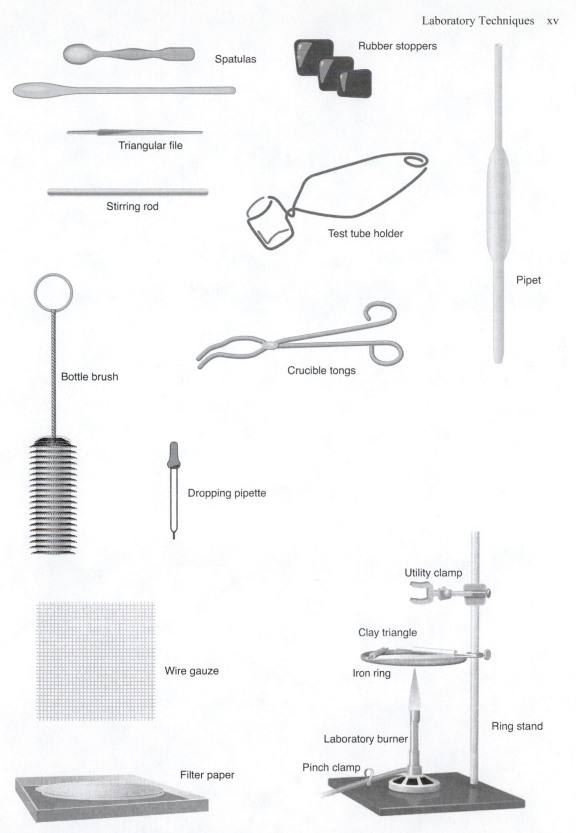

Spatulas

Rubber stoppers

Triangular file

Stirring rod

Test tube holder

Pipet

Bottle brush

Crucible tongs

Dropping pipette

Wire gauze

Utility clamp

Clay triangle

Iron ring

Ring stand

Laboratory burner

Pinch clamp

Filter paper

▲ **LT-3:** Laboratory Apparatus

Thinking Metric | 1

OBJECTIVES

1. To become familiar with the metric system of measurement—to "think metric."
2. To develop skills in metric measurement of length, volume, mass, and density.
3. To be able to convert from the English to the metric system and vice versa.

DISCUSSION

Some years ago we believed that the metric system, used by virtually all nations except the United States and a couple of small countries, would soon be adopted by even these holdouts. Certainly it would have facilitated trade among the nations. However, one of the laws of physics, the law of inertia, seems to have a counterpart in public affairs. Such change meets resistance. Some progress has been made in the United States, however, since we now see both English and metric units on some grocery and other items, and both units on many road signs. We have started to buy beverages by the liter. Almost everyone with a college education learns the metric system, but the general public is not yet accepting metrics in daily life and commerce. The United States economy is so large and dominant that it is simply not forced to conform to the rest of the world.

Metric units presently are used in all scientific work. If Thomas Jefferson's views had prevailed at the time of the nation's birth, the United States would have gone metric two centuries ago. Perhaps it is appropriate that Congress, in the bicentennial year, 1976, enacted legislation designed to eventually move the country into using the metric system, but the method adopted by the United States was persuasive rather than mandatory. Progress has been slower than expected, probably because of natural resistance to change and the cost of conversion. It will be difficult for the United States to remain different from the rest of the world and still be a major force in world trade. Certainly anyone who desires to be a global citizen will need to understand and use metric terminology.

Making metric measurements is incredibly simple by comparison with the traditional English system. Do not be overly concerned, initially, with converting from one system to the other. Focus your attention on metric units and you will soon be "thinking metric." After you are thoroughly familiar with metric units, the conversions will come somewhat naturally. First it is necessary to learn the prefixes that can be used with any of the basic units of measurement. Note the convenient multiples of 10. (In table 1.1 the most commonly used prefixes are indicated by boldface type.)

TABLE 1.1 Prefixes for the Basic Metric Units of Measurement

Exponential Expression	Decimal Equivalent	Prefix	Example
10^9	1 000 000 000	Giga-	Gigawatt
10^6	1 000 000	**Mega-**	Megaton
10^3	1 000	**Kilo-**	Kilogram
10^2	100	Hecto-	Hectare (of land)
10	10	Deka- (or deca-)	Decalog
10^0	1		Basic Unit
10^{-1}	0.1	Deci-	Decibel (sound)
10^{-2}	0.01	**Centi-**	Centimeter
10^{-3}	0.001	**Milli-**	Milliliter
10^{-6}	0.000 001	**Micro-**	Micrometer
10^{-9}	0.000 000 001	Nano-	Nanosecond
10^{-12}	0.000 000 000 001	Pico-	Picometer

Measurement of Length

The standard unit of length is the meter (m). A meter is 39.37 inches—a little over a yard. A person 5 ft. 10 in. tall measures 1.78 m or 178 centimeters (cm). (Note that in the metric system one unit of measure can be converted to another just by moving the decimal point.) Large distances, such as those between cities, are measured in kilometers (km). One kilometer is about five-eighths of a mile or about five city blocks. Highway signs reading "88 kilometers per hour" have the same meaning as "55 miles per hour" signs. Centimeters (cm) and millimeters (mm) are for measuring small objects like sheets of paper and machine parts. An ordinary piece of chalk is about 1 cm in diameter.

Extremely small dimensions are measured in special units. Dust particles and microscopic objects might be measured in micrometers (μm). Sizes of atoms or molecules are commonly expressed as picometers (pm). For example, a sodium atom is reported to have a radius of 186 pm, whereas a sodium ion (a sodium atom with its outer electron removed) has a 95-pm radius.

Area measurement introduces nothing really new. Square centimeters (cm^2), square meters (m^2), and square kilometers (km^2) are commonly used. Volume measurement, except for liquid volume, is similarly expressed as cubic centimeters (cm^3 or cc) or cubic meters (m^3). Since there are 100 cm in a meter, there are 100^3 or 1,000,000 cm^3 in a cubic meter.

Measurement of Liquid Volume

Imagine a container of milk just a little larger than a quart. This is a liter (L) or 1.06 qt. Liters are convenient to use in place of pints, quarts, or gallons. You will use liters and milliliters (mL) exclusively in this course where liquid volumes are required. Milliliters are for laboratory measurements. A teaspoon is about 5 mL, and a cup, which is a fourth of a quart, is then a little less than one-fourth of a liter. One cup equals 236 mL or 0.236 L.

A simple and important relationship exists between liquid volume and cubic volume. The liter was designed to equal exactly 1 cubic decimeter (dm^3) or 1000 cm^3. Actually they differ by a tiny amount, but for all practical purposes we can safely use the identity: $1\,cm^3 = 1\,mL$.

Measurement of Mass

Think of a kilogram (kg) of mass as a little over 2 lb (it's actually 2.20 lb).[1] The kilogram is appropriate for weighing items in a grocery store, sacks of fertilizer, people, or any item larger than a handful. Large items, such as shiploads of wheat, are sold by the metric tonne (t). A metric tonne is 1000 kg or 2200 lb, larger than an ordinary ton.

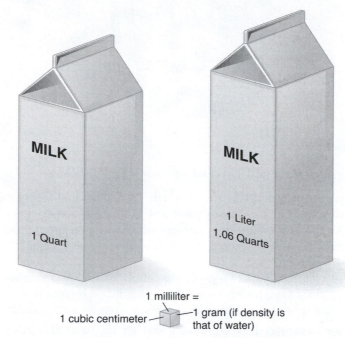

▲ **FIGURE 1.1** Metric volume measurement. Notice that 1 g of water occupies 1 mL (or 1 cm^3); so its density is 1 g/mL. Therefore, 1 kg (1000 g) of water is the same as 1 L (1000 mL) of water.

In the laboratory you will use grams (g) almost exclusively when weighing. A nickel coin has a mass of about 5 g. For small and precise quantities milligrams (mg) are appropriate. Good laboratory balances read in grams and are accurate to a milligram or even to a tenth of a milligram.

Measurement of Density

When both the volume and the mass of a substance are known, its density can be calculated. Density is defined as mass per unit volume.

$$\text{Density} = \frac{\text{mass}}{\text{volume}}$$

Any units of mass or volume may be employed to express density, but metric units are especially advantageous. One reason for this is that 1 mL of water weighs 1 g. You can easily summarize this convenient relationship.

---

[1]Weight is the pulling force of gravity upon a mass. When you weigh an object, you are really comparing its mass with a known mass. For most purposes on the Earth's surface, it is not necessary to make a distinction between mass and weight.

▲ **FIGURE 1.2** Common household metric measurements.

$$1 \text{ g of water (at } 4°C) = 1 \text{ mL of water} = 1 \text{ cm}^3 \text{ of water}$$

It follows that water has the convenient density of exactly 1.000 g per milliliter (g/mL) at 4°C. Therefore, when one knows the volume of a sample of water at 4°C, its mass is also known because they are numerically the same. At all temperatures near ambient (room) temperature the density is so nearly 1.00 g/mL that any difference is usually ignored. A 25-mL sample of water weighs 25 g. A 600-g mass of water occupies 600 mL or 0.6 L.

Since water has this special density of 1.00 g/mL, the density of any other substance in metric units reveals directly how many times more (or less) dense than water that substance is. This ratio, called *specific gravity* (a dimensionless quantity), has great usefulness in mass-volume calculations. Consider the following example: It was determined experimentally that an ice cube of exactly 10 cm^3 volume weighed 9.17 g/10.0 cm^3 or 0.917 g/cm^3. This tells us that ice is less dense than water (0.917 times or 91.7% as dense) and should float on water. A cubic foot of water is known to weigh 62.4 lb, so a cubic foot of ice should weigh 0.917 × 62.4 or 57.2 lb.

 ENVIRONMENT, CULTURE, AND CHEMISTRY

Measuring Tiny Amounts

For environmental protection and safety, one is often faced with measuring very tiny amounts. Recommended dosages are often expressed as micrograms (mcg) or milligrams (mg) per (/) kilogram (kg) of body weight. Contaminants are usually represented as parts per thousand (ppt), million (ppm), or billion (ppb). When discussing tolerances for toxins and contaminants, the person who thinks superficially is often tempted to demand zero contamination, forgetting that it is impossible to reach zero if one considers the last few molecules. In a cleanup operation, if 90% is removed on the first try, and then the 90% removal process is repeated over and over, it would still be impossible with any feasible number of repetitions to remove the last molecule. Similar problems exist in many areas: obtaining a perfect vacuum, reaching absolute zero, or having zero radiation (nature always provides some background radiation). Therefore, environmental regulations are stated in terms of some low level of contamination that can be safely tolerated. Remember, one molecule of any toxin or contaminant would certainly do no noticeable harm. Understanding this concept is increasingly important since our ability to detect smaller and smaller quantities is constantly improving to the point where it would become absurd to insist on finding no contaminant molecules. Projecting this ability into the future, one could foresee the possibility of detecting almost anything in something.

Pre-Laboratory Questions | 1

1. What are some arguments you can suggest for and against adoption of the metric system in the United States?

2. Arrange the following in order of decreasing length: meter, mile, centimeter, micron, picometer, foot, inch, kilometer.

3. Arrange the following in order of decreasing volume: cubic foot, gallon, liter, barrel, milliliter, cubic millimeter.

4. What can you say about the density of any object that can float on water?

5. Is the density of the human body greater than or less than 1.0 g/cm^3? Explain.

 **EXPERIMENT**

You are urged, in this experiment, to think metrically, but when you need to convert from one system to another, some useful conversion factors are shown in table 1.2.

TABLE 1.2 Some Useful Conversion Factors

Length	Weight	Volume
1 inch = 2.54 centimeters	1 pound = 454 grams	1 quart = 946 milliliters
1 yard = 0.914 meter	2.2 pounds = 1 kilogram	1 pint = 473 milliliters
39.4 inches = 1 meter	1 ounce = 28.35 grams	1 cup = 236 milliliters
0.62 mile = 1 kilometer		1.06 quarts = 1 liter
		1 cubic foot = 28.3 liters

1. Measurement of Length

Examine a meter stick. Notice the length of a meter, centimeter, and millimeter (the decimeter is of lesser importance). Measure the total length and width of the laboratory desk, using decimals for fractional units. For example, if the length is 3 m 54 cm and 4 mm, write it as 3.544 m or 354.4 cm. (Answer Questions 1a and 1b on the Observations and Results sheet.)

Measure the length and width, in centimeters, of an ordinary 8 1/2 × 11-in. sheet of paper and calculate its area in square centimeters. (Do your measured values agree with the statement 2.54 cm = 1 in.?) (Answer Questions 1c and 1d.) Stand against a wall and have your lab partner measure your height both in centimeters and in meters. Measure your partner's height. Now that you are familiar with centimeters, estimate and then measure your shoe length and your neck size. (Answer Questions 1e, 1f, 1g, and 1h.)

2. Measurement of Volume

Measure in centimeters the dimensions of a rectangular object such as a block of wood or a box of filter paper. Multiply its length, width, and height together to calculate its volume in cubic centimeters. (Answer Questions 2a and 2b.) Roughly *estimate* the volume of a piece of new chalk in cubic centimeters.[2] (Answer Question 2c.) Fill a test tube[3] to the brim with water. Determine the volume of the water by pouring it into a graduated cylinder. The cylinder is calibrated in such a way that you should read the volume *under* the meniscus (figure 1.3). Use this water in the next part of the experiment. (Answer Questions 2d and 2e.)

---

[2]You could estimate the volume by assuming the chalk to be a rectangular solid, or you could use the formula for the volume of a cylinder, $V = \pi r^2 h$.

[3]Throughout this manual, 10 cm test tubes are recommended unless another size is indicated.

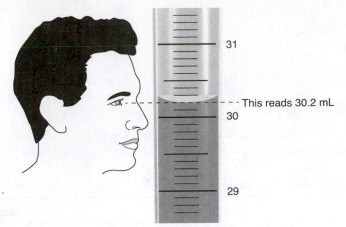

This reads 30.2 mL

▲ **FIGURE 1.3** Proper sighting in volume measurement. Hold the meniscus (the thin opaque crescent at the surface) level with your eye and read the volume under it.

3. Measurement of Mass

Weigh the water whose volume you measured in part 2. Most balances can be tared to subtract the weight of the container, so only one weighing is required (Answer Questions 3a and 3b.) Weigh a nickel coin and record its mass. (Answer Questions 3c and 3d.) Fold an 8 1/2 × 11-in. sheet of paper several times and weigh it. Estimate and then calculate what a 10 × 10-cm piece of the same paper should weigh. (Answer Questions 3e and 3f.)

Determine the mass of a tiny object, such as an aspirin tablet, by a method you can use even if your balance is unable to give sensitive readings. Count out 100 tablets, place them in a tared container (in a dish or on a piece of paper), and weigh the tablets. Divide by 100 to get the weight (mass) of 1 tablet. This is the principle by which masses of atoms and molecules are determined. (Other suitable objects are saccharin tablets, toothpicks, paper clips, postage stamps, or grains of wheat. If wheat is used you will be calculating the value of a rarely used unit of mass called the grain, about one-sixteenth of a gram.) (Answer Questions 3h, 3i, 3j, 3k, and 3l.)

4. Determination of Density

A. Density of a Rectangular Object

Since density is mass per (divided by) unit volume, we need both mass and volume. You have already determined the volume of the rectangular object used in part 2. Weigh the object to complete the measurements needed for the calculation of its density. (Answer all parts of Question 4A.)

B. Density of an Irregularly Shaped Solid

Now determine the density of a small, irregular shaped object, such as rocks, a rubber stopper or a piece of metal. Weighing is not a problem, but measuring the volume may take a little thought. Propose a method and secure the approval of your instructor. (Hint: Remember for water, $1\,mL = 1\,cm^3$.) Refer to figure 1.4. (Perform the measurement and answer all parts of Question 4B.)

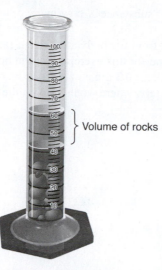

} Volume of rocks

▲ **FIGURE 1.4** Measuring the volume of irregularly shaped objects.

QUESTIONS AND PROBLEMS

1. An automobile gasoline tank holds 24 gal. If the tank were filled in Paris, France, how many liters would it take?

2. The road signs along the superhighway to Mexico City set the speed limit at 110 km/hr. What would be the reading in miles per hour on your speedometer?

3. Find the volume in cubic centimeters of a box 25 mm × 12 cm × 0.20 m. What is the maximum number of grams of water that the box can hold?

4. The unit of length in use for extremely large astronomical distances is the light-year, the distance light can travel in one year. The nearest star (other than our own sun) is 4 light-years away. If the speed of light is 186,300 miles per second, how far away is that star in miles?

5. Calculate your weight in kilograms. How would you find the density of your body?

6. Suppose you were to propose a new unit of time called the *centiday*. What would be its length in minutes? What would be the length of a *kiloday* in years?

7. Many substances such as the familiar aspirin, sleeping pills, and ammonia gas are toxic in large quantities, but benign or even beneficial in small amounts. Suppose substance X was known to cause harm if inhaled for a prolonged time at a concentration of one part per thousand (1 ppt) in air. Extensive testing had indicated that concentrations of 1% of that, 10 parts per million (10 ppm), caused no harm, and to be conservative OSHA had set the industrial standard at 3 ppm. Now suppose in your work environment the air has been analyzed and found to contain 1 ppm of X.
   a. Are you safe?
   b. If you take 0.5 liter of air into your lungs every second, how many liters of air would you inhale in 8 hours?
   c. Air weighs approximately 1.2 grams/liter. How many grams, and kilograms of air would you inhale in 8 hours?

d. If that air contains 1 ppm of substance X how many grams and milligrams of X would you inhale in that time?

e. Different toxins are purged from the body at different rates. Still remembering that X is hypothetical, let us further assume for this exercise that the body, through normal intake and excretions, is able to flush out at least 0.3 gram of X from the body daily. Based on your calculations and allowing a margin of safety, where would you set the industrial standard?

EXPERIMENT

Thinking Metric | 1

OBSERVATIONS AND RESULTS

1. Measurement of Length

a. Length of laboratory desk _____ cm; _____ m

b. Width of laboratory desk _____ cm; _____ m

c. Dimensions of sheet of paper _____ cm × _____ cm

(Does your calculation agree with

1 in. = 2.54 cm?)

d. Area of sheet of paper _____ cm²

e. Your height _____ cm; _____ m

f. Your partner's height _____ cm; _____ m

	Estimated	*Actual*

g. Shoe length _____ cm; _____ cm

h. Neck size _____ cm; _____ cm

2. Measurement of Volume

a. Dimensions of rectangular solid _____ cm × _____ cm × _____ cm

b. Volume of rectangular solid _____ cm³

c. Estimated volume of a piece of chalk _____ cm³

d. Volume of water in test tube _____ mL

e. How many grams should the water weigh? _____ g

3. Measurement of Mass

a. Weight of water _____ g

b. How closely do the mass and volume of water that you determined
correspond to 1 mL = 1 g?

c. Weight of nickel coin _____ g

d. In view of what you have just done, suggest a way to communicate to a child the approximate size
of a gram.

e. Weight of 8 1/2 × 11-in. sheet of paper _____ g

f. Your estimate of the weight of a 10 × 10-cm piece of paper _____ g

g. Calculated weight of a 10 × 10-cm piece of paper _____ g

h. Weight of 100 tablets (or other small objects) _____ g

i. Weight of 1 tablet (weight of 100 tablets divided by 100) _____ g

j. Is this an actual or an average weight? _____ g

k. Would you classify the objects in your sample as quite uniform in mass?

l. Would you say this weight is representative of one of the objects you might pick up at random?

4. Determination of Density

A. *Density of a Rectangular Object*

Volume of rectangular object (from part 2) _____ cm^3

Weight of rectangular object _____ g

Calculated density of object (show calculations) _____ g/cm^3

Would these samples have the same or nearly the same density: (a) different sized samples of the same kind of wood and (b) samples of different kinds of wood?

Considering that wood floats on water, what can you say about the relative densities of wood and water?

Would a piece of wood float on mercury (the density of mercury is 13.6 g/mL)? Explain.

B. *Density of an Irregularly Shaped Object*

Weight of object _____ g

Volume of object _____ mL or cm^3

Calculated density of object _____ g/mL or g/cm^3

How many times denser is the irregularly shaped object than water? (Example: If x has a density of 1 g/cm^3 and y has a density of 2.5 g/cm^3, y is 2.5 times denser than x.)

If water has a density of 62.4 lb/ft^3, how many pounds would a solid cubic foot of the irregularly shaped object weigh (no air spaces)?

Molecular Motion: The Particulate Nature of Matter | 2

OBJECTIVES

1. To develop the concept of matter as particles—the way chemists think of matter.

2. To examine some of the evidence for the discontinuity of matter and for particle motion in matter.

3. To learn how the theory of molecular motion explains some familiar phenomena in gases, liquids, and solids.

4. To gain an appreciation of the relative speeds of molecules of different masses.

DISCUSSION

Questions about the nature of matter have captured the attention of thinkers since ancient times. Some early philosophers believed it was theoretically possible to subdivide matter into smaller and smaller pieces indefinitely (continuous matter). Others championed the view that, ultimately, matter consisted of tiny particles which could not be further divided (discontinuous matter). For centuries a fundamental argument centered on this point—whether matter is continuous or discontinuous. The question is not debated today because there now exists an enormous body of evidence supporting the view that matter is discontinuous.

Particle Theory of Matter

Proof of the existence of particles is perhaps best left for sophisticated experiments in nuclear and particle physics. What we shall do today is think about how a theory postulating particles in motion can account for all of the observed properties and behavior of gases, liquids, and solids. In the experimental exercises you should decide whether a theory of molecules (or ions) in motion suitably explains what you observe. Usually, in scientific work, the simplest explanation which accounts for *all* of the observed phenomena is accepted as the correct one.

Based on modern evidence, we can construct models of matter (graphical pictures, mental images, or even mathematical models) which help us to understand what gases, liquids, and solids are like. According to such models, if a substance were subdivided many times, ultimately minute unit particles, called molecules (for covalent compounds), ions (for ionic compounds), or atoms (for elements), would be obtained. These unimaginably small particles cannot be further subdivided and still *retain the identity* of the substance. Therefore, the term molecule (or ion or atom) will be used to designate the basic particle of a substance.

Gases

According to the moving-particle theory (called the kinetic-molecular theory by scientists), a gas consists of molecules moving furiously in random directions with relatively vast empty spaces separating them. The average speed of the molecules is directly related to absolute (Kelvin) temperature.[1] Consider a balloon full of a familiar gas such as oxygen, O_2. A typical oxygen molecule darts about, crashing into its neighbors and against its confining walls with roughly the velocity of a speeding bullet. If the oxygen is heated, its molecules move still faster. If the balloon is squeezed to a smaller volume without releasing any gas, the molecules themselves do not get smaller, the space between them does, and therefore they strike each other and the walls more frequently. The striking of the confining walls by the molecules produces the pressure of the gas.

Liquids

The molecules of a liquid also are believed to be moving randomly, but in close contact with each other. Thus, in contrast to a gas, a liquid has a definite volume and is practically incompressible. Molecular motion accounts for the fact that liquids evaporate. Every liquid has a certain fraction of its molecules moving upward fast enough to break through the surface (evaporate) into the space above.

When the high-speed molecules leave the liquid, the average speed of those remaining is slower, and the temperature of the liquid is thereby reduced by evaporation. (The average speed of molecules in a liquid, like the speed of molecules in a gas, is related to the temperature.) When leaving the liquid surface, molecules are slowed (cooled) because energy (heat of vaporization—see experiment 4) must be supplied to overcome the attraction between neighboring molecules. Thus, the evaporation of a liquid has a cooling effect. Evaporating water is an effective cooling method in desert climates. Similarly, perspiring is the body's way of cooling off.

Not all liquids have the same tendency to evaporate. Liquids whose molecules have lighter masses or weaker attractions for each other are more easily vaporized. Higher temperatures also encourage more molecules to escape into the vapor. For each liquid there is a temperature at which the escaping tendency becomes so great that the liquid boils. Condensation occurs when molecules of the vapor attract each other sufficiently to coalesce into droplets. This happens when they are cooled (slowed) or compressed (crowded together).

Solids

According to our theory, particles in a solid are also in motion. Like liquid particles, they are essentially touching each other, but unlike liquid particles, they are arranged in regular three-dimensional patterns so that each molecule can vibrate only in the space bounded by its nearest neighbors. This explains why solids maintain their shapes and are essentially incompressible. Diffusion (intermingling) is not possible in solids unless there happens to be some empty spaces or defects in the pattern. Solids do vaporize to a certain extent because some of the molecules at the surface vibrate rapidly enough to escape. Melting is the complete breakdown of the ordered pattern of a solid to the randomness of a liquid. Rather than melting, some solids, such as dry ice (solid carbon dioxide, CO_2) go directly from solid to vapor. These are said to *sublime*.

When a solid dissolves in a liquid, the particles (molecules or ions) immediately begin to move throughout the solution. When a colored substance is dissolved in water, one can see the leading edge of the cloud of particles moving through the water. During today's experiment, try to visualize the movement of the colored particles as they progress.

---

[1]The average molecular speed squared is directly proportional to the absolute temperature and inversely proportional to the mass.

Diffusion

Gases diffuse into each other completely, because of the random motion of their molecules and the relatively large spaces between them. A sample of perfume released in a corner of a room soon will completely fill the room by mingling with the molecules of air. You will become familiar with a number of odors in this course. When you detect an odor, remember that molecules of that substance have traveled in a zigzag path due to billions of collisions along the way with other molecules in the air. If the odorous material is a liquid (such as perfume) or a solid (like a mothball), it must first vaporize to a gas before it can migrate and strike the receptors in your nose. Sight is different. You can see perfume in the bottle without having perfume molecules coming toward you (light travels independently of particles), but you cannot smell perfume without encountering perfume molecules in your nostrils. Also, remember that in a substance which produces an odor, the source is being used up, so the odor will not persist forever.

Not all molecules move at the same speed, even at the same temperature. As might be expected, heavier molecules move more slowly and lighter ones move faster. For example, hydrogen molecules, H_2 (molecular mass = 2), move faster than oxygen molecules, O_2 (molecular mass = 32).

ENVIRONMENT, CULTURE, AND CHEMISTRY

Atoms, Molecules, Ions, and Nature's Cycles

As shown in this experiment, atoms, molecules, and ions are in motion when a gas or in solution. Our environment involves these particles of nature in many ancient cycles, many of which have been affected by humans. In the carbon cycle for example, carbon dioxide, CO_2, is converted by photosynthesis in plants to carbohydrates which are then burned or metabolized, producing CO_2 again. Use and regeneration of carbon, in CO_2 and carbohydrate molecules, and oxygen, in the O_2 molecule, are part of this cycle. The atmosphere surrounding our planet contains a huge amount of CO_2 and O_2, but if one could have followed a particular carbon or oxygen atom since the beginning of humans (approximately 200,000 years), it is almost certain that the same atom has been used many times in the carbon cycle. Humans are adding significant quantities of CO_2 to the atmosphere by burning fossil fuels and coal for heating homes, generating electricity, and producing power for motor vehicles; this has severely unbalanced the carbon cycle and probably caused global warming. Nature and living species have several nitrogen cycles. One of these is the nitrogen, N_2, fixation cycle, which is called this because the nitrogen molecule, N_2, in the atmosphere is converted (fixed) by bacteria in plants called legumes to give the ammonium ion, NH_4^+, and other nitrogen particles. In turn, other bacteria in the soil can reverse the cycle by digesting animal waste materials, such as urea, NH_2CONH_2, regenerating nitrogen, N_2. The nitrogen cycle is also out of balance because of the large amounts of nitrogen fertilizers (liquid ammonia, NH_3, and NH_4^+ salts) that are used by farmers. Some of the fertilizer is washed into streams, rivers, and lakes, enriching the water, leading to excessive plant and algae growth, and depleting the water of O_2.

The water, H_2O, cycle is of major importance. Water is so much a part of our daily lives that it is easily taken for granted. H_2O molecules evaporate from the ocean, lakes, rivers, and even the ground, forming clouds, which leads to precipitation onto the earth again as rain or snow—truly an ancient cycle. Furthermore, water percolates through the earth to form aquifers, large bodies of fresh water deep under the ground; these aquifers have become a major, direct source of fresh water. Humans have interfered with the water cycle in several ways. Because of overpopulation and the reliance on single-family houses, more fresh water is being used than is produced by the ancient cycle. Our "used" water too often contaminates the fresh water, most commonly by under treatment and insufficient purification or by direct runoff into the oceans, lakes, and rivers. Contamination extends even to the deep aquifers, as was recently reported in the detection of t-butyl methyl ether, used to increase octane in gasoline, and the serious carcinogen chromium-six, Cr^{+6}. Remember the Erin Brockovitch lawsuit/movie; the authorities promised that the Cr^{+6} in the surface soil would never reach the aquifer below. Recently it did. This serious fresh water problem can easily become a catastrophe unless the state and federal authorities begin to manage it with insight and care.

Name _____

Date _____ Lab Section _____

Pre-Laboratory Questions | 2

1. Explain how it is possible for you to be aware, when blindfolded, that a person wearing perfume has entered the room.

2. What can you say about the average distance between molecules in a gas as compared with that in a liquid?

3. Make a statement describing the way molecules are arranged in solids and in liquids.

4. Would molecules be expected to move faster or slower as the temperature is increased?

5. Would chlorine, Cl_2, molecules (molecular mass = 71) be expected to move faster or slower than hydrogen, H_2, molecules (molecular mass = 2)?

6. Experimental evidence indicates that ammonia molecules travel about 2000 ft/second at room temperature. Explain, using the idea of particle motion, why it takes several seconds for the odor to travel a few feet.

EXPERIMENT

1. Diffusion of Gases

A. Diffusion of Methyl Salicylate (a Group Experiment)

In this part of the experiment your instructor will release a vapor and you will determine how much time elapses before you detect the odor of the substance. The substance to be used is methyl salicylate—also called oil of wintergreen—$C_8H_8O_3$ (molecular mass = 152). Methyl salicylate is an important ingredient in liniments. When it is rubbed on the surface of the skin, it causes a mild inflammation which increases the blood supply to the area and relieves sore muscles.

This portion of the experiment needs to be done first, before the room is filled with other odors. After making sure that windows are closed and fume hoods turned off so there is no draft, the instructor will vaporize methyl salicylate by quickly pouring about 0.5 mL of the liquid into a watch glass on a heated hot plate. The hot plate should be placed in a central position in the room. Note exactly when the vapor is released and determine the time required for the vapor molecules to reach you. You should not be standing more than 5 or 6 m from the point of release, and, preferably, there should be no one standing between you and the source of the vapor. Do not move until all of the students have detected the odor. Mark the spot where you are standing with a piece of chalk, and determine the distance from the hot plate to the chalk mark by stretching a piece of string across this distance. Then measure the length of the string. Record the data and answer the questions. (Answer Question 1A, 1–7 on the Observations and Results sheet.)

B. Diffusion of Ammonia and Hydrogen Chloride

DEMONSTRATION: Your instructor will remove the stoppers from bottles of concentrated hydrochloric acid, HCl, and concentrated ammonia, NH_3, solution (alternatively labeled ammonium hydroxide, NH_4OH). Molecules of HCl and NH_3 vaporize from their concentrated solutions and produce the odors. Watch as the mouths of the bottles are brought close together and observe the result.[2] (Answer Question 1B, 1–2.)

C. Comparison of the Diffusion Speeds of Ammonia and Hydrogen Chloride

DEMONSTRATION: Your instructor will assemble a long *dry* tube (about 60 × 2 cm) with a cotton plug at each end (figure 2.1). Simultaneously moisten the plugs with 10 to 15 drops (from a dropping pipet) of concentrated hydrochloric acid, HCl, at one end, and concentrated ammonia, NH_3, solution (ammonium hydroxide, NH_4OH) at the other. Insert the rubber stoppers. Wait until a cloud forms inside the tube. Note and record its distance from each end. (Answer Question 1C, 1–5.)

---

[2]In all laboratory work be careful to keep chemicals off your skin and, especially, out of your eyes. If you get a chemical on you, wash it off immediately. Eyes should be irrigated thoroughly at an eyewash fountain.

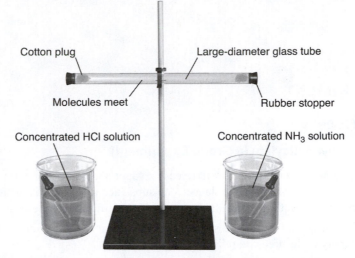

▲ FIGURE 2.1 Apparatus for comparing the speeds of HCl and NH_3 molecules.

2. Evaporation of Liquids

Pour 1 mL of acetone onto a watch glass under the hood. Move the acetone back and forth on the watch glass and note how easily it evaporates. With your hand, check the temperature of the bottom of the watch glass as the acetone evaporates. Boil 25 mL of water in a beaker and determine its boiling temperature with a thermometer. Cool the water and add enough salt (sodium chloride, NaCl) to make the solution about 15% to 20% salt by weight (about 3 to 5 g of salt). Heat the salt solution, and again note the boiling temperature. Record both boiling temperatures. Is the boiling point of the solution different from that of the pure solvent? (Answer Question 2, a–f.)

> **Pour the salt water down the drain.**

3. Vaporization of Solids

Sniff some paradichlorobenzene, $C_6H_4Cl_2$, which is identical to commercial mothballs. Consider the process whereby your nose receives the odor from the mothballs.

Put one small (pea-sized) piece of solid iodine in an Erlenmeyer flask, and place the flask on a wire gauze on your ring stand (figure 2.2). Heat gently until sublimation takes place. This is one of the few colored vapors. Are molecules in motion? Continue heating, forcing as little vapor as possible from the mouth of the flask. Note any condensation of iodine vapor on the inside of the flask. (Answer Question 3, a–c.)

> **Dispose of the iodine in the appropriate container.**

4. Diffusion in Solutions

Place a 100-mL beaker, containing water to a depth of about 1 cm, on a piece of lined white paper. Determine the temperature of the water; then allow the water to become very still. Using a forceps, remove several crystals of potassium permanganate, $KMnO_4$, from the reagent bottle, and carefully place them on the bottom of the beaker at a place where the lines are perpendicular to the beaker's edge (figure 2.3). Note the time required for the colored ions to travel 5 mm along a line toward the center. Rinse the beaker and again fill it with water to a depth of 1 cm. Heat the water to 70 or 80°C, record its temperature, and allow it to become still. Again add crystals of $KMnO_4$ in the same way, and note the time of migration for the same distance at the higher temperature. (Answer Question 4, a–g.)

Dispose of the $KMnO_4$ in the appropriate container.

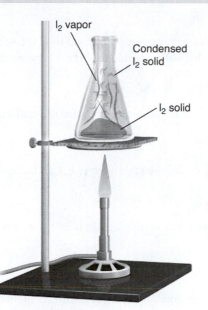

▲ **FIGURE 2.2** Apparatus for the sublimation of iodine.

▲ **FIGURE 2.3** Beaker placed on lined paper to show migration of ions.

QUESTIONS AND PROBLEMS

1.   Copper sulfate is blue, copper chloride is blue, potassium chloride is colorless, and sodium sulfate is colorless. What could you confidently predict about the color of potassium sulfate?

2.   Based on information in the previous question, and with the additional information that potassium permanganate is purple, what color would you expect for copper permanganate?

3.   Does hearing a sound involve an encounter of molecules with the eardrum? Explain.

4.   a. Light reflecting off a gold coin reaches your eye. Can this continue indefinitely? Why?

     b. A dead rat in the rafters sends an unpleasant odor to your nose. Can this continue indefinitely? Explain.

5.   Let us suppose Julius Caesar drank a cup of coffee (0.25 liters of water) that fateful Ides of March morning in 44 B.C. Try to estimate how many of those water molecules, if any, were in your cup of coffee this morning. For simplicity you will need certain information and make several assumptions. There are about 6×10^{23} molecules (Avogadro's number) in 0.02 liters of water. The earth's surface (4000 miles diameter) is about three-fourths covered with water to an average depth of about 1.5 miles. Assume the waters of the earth get thoroughly mixed. (If it only got half mixed, your answer would double.) The formula for the surface area of a sphere is $4\pi r^2$. There are 5280 ft. in 1 mile (say 5000), and a cubic foot of water is about 30 liters. Remember, very large and very small numbers all follow the usual rules of mathematics, but it is easier to keep decimal points straight by using exponents. Your answer should be somewhere near half to twice the actual number. Consider how much water you and Caesar drink in a lifetime, would you say water is reused?

6.   To appreciate how tiny atoms and molecules are, do this exercise. Suppose someone made a teaspoonful (say 7×10^{23} atoms) of a rare precious metal and distributed it equally (submicroscopic portions) to all 7 billion inhabitants of the earth. Thinking it might be valuable some day you tuck your speck under your eyelid for safe keeping. Miraculously, someone offers one cent for every 10,000 atoms of this metal. This news brings tears of joy to your eyes and the speck is washed out. How much are you tears worth?

Name ______________________________

Date ______________ Lab Section ______________

EXPERIMENT

2

Molecular Motion: The Particulate Nature of Matter

OBSERVATIONS AND RESULTS

1. Diffusion of Gases

A. *Diffusion of Methyl Salicylate (a Group Experiment)*

1. Time required for odor to arrive ____________ seconds

2. Straight-line distance traveled by molecules ____________ cm

3. Apparent speed (rate of migration) of molecules (report as a decimal value) ____________ cm/second

4. Do the observations of this experiment suggest that molecules are in motion? Explain how the odor came to you from its source.

5. Try to explain how the odor arrived at your location if you assume matter is continuous (no particles are involved).

6. Would you expect ammonia gas (molecular mass = 17) to migrate faster or slower than methyl salicylate? Explain.

7. Suggest at least two explanations for the fact that considerably different speeds of diffusion were reported for oil of wintergreen by members of your class.

   B. *Diffusion of Ammonia and Hydrogen Chloride*

   1. What did you observe when ammonia and hydrochloric acid bottles were brought together?

   2. How does this experiment demonstrate that molecules travel through space (air)?

   C. *Comparison of the Diffusion Speeds of Ammonia and Hydrogen Chloride (a Demonstration)*

   1. Distance traveled by HCl molecules   _____ cm

   2. Distance traveled by NH_3 molecules   _____ cm

   3. Which molecules move faster?

   4. Which molecules have a lower mass?

   5. Did the lighter molecules move faster than the heavier ones?

2. Evaporation of Liquids

   a. Recalling your previous experiences with the evaporation of water, would you say that acetone molecules escape more or less readily from the liquid surface than do water molecules? How do you know?

   b. What do the relative rates of evaporation of these liquids tell you about the attractive forces between neighboring molecules in water and in acetone?

c. What happened to the temperature of the acetone as evaporation progressed? Explain this observation on the basis of the motion of molecules.

d. Boiling temperature of pure water   _____ °C

e. Boiling temperature of salt water   _____ °C

f. Suggest a reason why molecules of one solution might escape more readily (and thus boil at a lower temperature) than those from the other solution. (Hint: Consider deterrents to escape.)

3. Evaporation of a Solid

a. Explain how solid mothballs (paradichlorobenzene) activated your nose. Use the moving particle theory in your explanation.

b. Solid iodine is so dark that it appears black, but it is actually purple. What color is iodine in the gas phase? On the basis of these observations, what can you conclude about the identity of the molecules in the solid and gas phases?

c. In sublimation, molecules leave an ordered arrangement in the solid to take a random orientation in the vapor. Did you obtain evidence that iodine molecules in the vapor returned to an ordered pattern when cooled to a solid? Explain.

4. Diffusion in Solutions

a. What evidence did you observe for the movement of the particles (in this case, ions) of potassium permanganate, $KMnO_4$?

b. Time for $KMnO_4$ to travel 5 mm in cooler water

_____ seconds

c. Temperature of cooler water

_____ °C

d. Time for $KMnO_4$ to travel the same distance at

the higher temperature

_____ seconds

e. Temperature of warmer water

_____ °C

f. What do you conclude about the rate of particle (ion) movement as the temperature increases?

g. Realizing that every molecule or ion collides billions of times with neighboring particles as it proceeds in its random migration through a medium, sketch what might be a typical pathway for a particle.

What Is a Chemical Reaction? | 3

OBJECTIVES

1. To learn the difference between a chemical change and a physical change.
2. To acquire an understanding of what constitutes a chemical reaction.
3. To learn the meaning and significance of chemical equations.
4. To develop an awareness of bond breaking, bond making, and energy effects in chemical reactions.

DISCUSSION

The burning of gasoline, the souring of milk, the digestion of food, the freezing of water, and the dissolving of salt are familiar examples of changes in matter. The first three are chemical changes and the last two are physical changes. What is the difference? How are tiny atoms, molecules, and ions involved in these changes? The answers to such questions are the proper concern of chemistry, for chemistry is the study of matter and its transformations.

In a chemical change, new substances are formed which have properties completely different from those of the original starting materials. We may expect to see flames, gas bubbles, changing colors, different solubilities, and other evidence of chemical activity. In all cases of chemical change, substances disappear, and new ones appear. Other examples of chemical change are the corrosion of iron, the tarnishing of silver, and the cooking of food. Physical changes do not destroy the original substances or form new substances. Often a substance seems to disappear, but if there has been no chemical change it is recoverable. For example, when salt dissolves in water, it may be recovered easily by evaporating the water. Other examples of physical change are grinding a solid to a powder (the powder can be recompacted) and boiling water to form steam (the steam can be condensed).

Chemical Reactions and Equations

When we say a chemical change has occurred, we really mean a chemical reaction has taken place. The original substances (*reactants*) have interacted (reacted) to form new substances (*products*). A chemical reaction is most conveniently represented by an equation, a type of chemical shorthand. Equations describe reactions by employing symbols to represent elements, and groups of symbols (formulas) to represent compounds. Reactants, written on the left side of the equation, are changed into products, which are indicated on the right. An arrow indicates which way the reaction proceeds.

The reaction of hydrogen with chlorine to form hydrogen chloride may be stated in words, but an equation actually describes the reaction with greater clarity and economy.

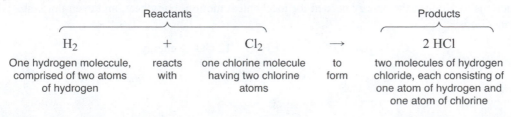

Reactants				Products
H_2	$+$	Cl_2	$\rightarrow$	$2\ HCl$
One hydrogen molecule, comprised of two atoms of hydrogen	reacts with	one chlorine molecule having two chlorine atoms	to form	two molecules of hydrogen chloride, each consisting of one atom of hydrogen and one atom of chlorine

The symbols H and Cl indicate that the elements hydrogen and chlorine are involved in the reaction. The formulas H_2, Cl_2, and HCl tell how many atoms of which kind are in the molecules of each substance. The integers (coefficients) in the equations, 2 before the HCl and 1 (understood) before the H_2 and Cl_2, indicate the number of molecules of each kind which are involved. Later (experiment 11) we shall see that an equation also reveals precise data about the relative quantities in which the various reactants and products enter the reaction. All chemical changes (reactions) can be described by chemical equations. In this experiment, you will begin a new experience—writing equations for simple reactions. Do not be discouraged if you make mistakes or if you have not learned the symbols for the elements. They will soon become familiar to you.

Chemical Bonds

Atoms combine with other atoms in many different ways to form molecules. Atoms in molecules are held together by attractive forces we call *bonds*. In a chemical reaction the disappearance of the reacting substances is the result of the dismantling of the original molecules. The appearance of new substances results from the reassembling of the atoms in a different way to form new molecules. The bonds which are broken or made in the reaction of hydrogen with chlorine may be represented schematically as

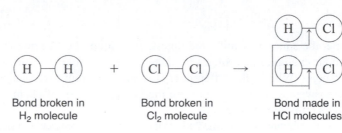

| Bond broken in H_2 molecule | Bond broken in Cl_2 molecule | Bond made in HCl molecules |

Energy and Chemical Reactions

Essentially all chemical reactions are characterized by a gain or loss of energy. Those reactions which give off heat energy are said to be *exothermic*, and those which require (absorb) heat are called *endothermic*. The burning of natural gas is a classic example of an exothermic reaction. Certain physical changes also absorb or release energy.

We can understand the exothermic and endothermic nature of chemical reactions by examining what happens to the bonds between atoms. The breaking of a bond requires energy because the attractive forces between the atoms must be overcome. The reverse is also true—energy is released when a bond is formed. Since chemical reactions involve the breaking and making of bonds, some energy must be supplied (for bond breaking) and some released (in bond formation) in the process. Whether a reaction is exothermic or endothermic depends upon the strength[1] and number of bonds involved. Consider again the reaction between hydrogen and chlorine. The reaction is known to be exothermic. This tells us that more energy is released in the formation of two H—Cl bonds than is required to break one H—H bond and one Cl—Cl bond. Reactions of molecules of much greater complexity, having many bonds, may be analyzed similarly. When table sugar (sucrose) is taken into the body it undergoes a series of complicated reactions to produce the energy needed for mechanical motion, body heat, and even thinking. The overall reaction is

$$C_{12}H_{22}O_{11} + 12\,O_2 \rightarrow 11\,H_2O + 12\,CO_2 + \text{energy.}$$

---

[1]The strength of a bond is the energy required to break it or the amount of energy released when it is formed.

In the metabolic reactions in the cells, many bonds are broken (sugar and oxygen molecules come apart), and new bonds are formed (in water and carbon dioxide). It turns out that more energy is released in bond formation than is required for bond breaking, so the reaction is exothermic, giving life-sustaining energy.

Throughout this course you will make numerous observations and measurements on the macro (normal laboratory size) scale of operation. In addition to understanding what is happening at this level, where you can see and manipulate things, always think of what is happening to the tiny atoms and molecules on the micro scale. When you observe a chemical reaction, try to imagine molecules smashing into one another and coming apart to form new molecules. Imagine bonds being broken and reformed. Develop in your mind's eye a particulate view of matter. Remember, the dimensions of these particles are inconceivably small (around 10^{-8} cm); yet this is the level where the real action is.

ENVIRONMENT, CULTURE, AND CHEMISTRY
Nature's Two Fundamental Reactions

Two fundamental, basic reactions occur in our environment. They are both complex, involving many reactions in the overall reaction, and are opposite in direction. One of these reactions will be studied in today's experiment, the combustion (oxidation) of carbohydrates in the body to produce carbon dioxide, water, and energy, including heat to keep the body at constant temperature. This combustion reaction occurs in nature as burning, ignited by heat or a flame, or by catalysis by enzymes in living creatures of all types. The other reaction, called photosynthesis, goes in exactly the reverse direction, using carbon dioxide from the atmosphere and energy from the sun to produce carbon compounds such as carbohydrates and cellulose. This reaction occurs in all plants and trees. Therefore, in a cyclic process, carbon dioxide produced in the first reaction process is converted back ("fixed", reduced) to carbohydrates and cellulose by the second reaction. The photosynthesis process also involves a complex set of reactions, requiring the catalyst chlorophyll—a green molecule, which gives the distinctive color to plants, trees, and algae—to capture the sun's energy. Obviously, a balance must be maintained in these two fundamental, natural reactions. Currently, with the enormous burning of fossil fuels, combustion is gaining and the level of carbon dioxide is rising, as will be discussed in Experiment 19.

Name _____

Date _____ Lab Section _____

Pre-Laboratory Questions | 3

1. In the souring of milk, what are some of the reasons for believing that a chemical change, rather than a physical change, has occurred?

2. Tell which of the following are chemical changes: (a) fermenting sugar, (b) winding a clock spring, (c) burning wood, (d) evaporating ether, (e) cooking meat, (f) dissolving sugar, (g) mixing sand and salt, and (h) rusting iron.

3. What happens to the reacting substances in chemical reactions?

4. Give one reason for writing chemical equations. Make a distinction between a chemical reaction and a chemical equation.

5. What is the difference between an endothermic reaction and an exothermic reaction?

6. Give a simple definition of a chemical bond. What happens to bonds between atoms in a chemical reaction?

EXPERIMENT

1. Iron and Sulfur

In this part of the experiment you will examine the properties of two substances, which we have not discussed, iron and sulfur. Then you will mix them and decide, on the basis of observations and tests, whether a chemical reaction has occurred. Finally, you will heat the mixture and see if iron and sulfur react at the higher temperature.

A. Properties of Iron and Sulfur

Obtain some *reagent* iron *dust* and some sulfur, each about the size of a pea. Note the physical properties of each element—its color, hardness, and so forth—and make the following tests.

1. Find out how each substance is affected by a magnet. Place the materials on a sheet of paper and move a magnet underneath. (Answer Question 1A, 1–2 on the Observations and Results sheet.)

2. Put the sulfur and iron samples from question A.1 in separate test tubes and add dropwise sufficient dilute (6 M) hydrochloric acid, HCl, to cover the solids. Note whether a chemical reaction takes place with either iron or sulfur. Do you smell any gaseous products? When the iron and acid are mixed, you may observe a faint smell of hydrogen sulfide because even reagent iron dust is often contaminated with iron sulfide. This odor does not result from a reaction of iron and acid. (Answer Question 1A, 3–4.)

3. DEMONSTRATION: Your instructor will mix by stirring small amounts of iron and sulfur with carbon disulfide, CS_2, in separate test tubes under the hood. Is either iron or sulfur affected by the CS_2? Transfer each *liquid* with a dropper onto a separate watch glass under the hood so that the CS_2 can evaporate. What do you observe on the watch glasses after evaporation? Was the change physical or chemical? (Answer Question 1A, 5–7.)

Dispose of all waste in the appropriate containers.

B. Do Iron and Sulfur React?

DEMONSTRATION: Your instructor will ask each of you to connect your Bunsen burner to the gas supply with a piece of gas tubing, show you how to light the burner, and adjust it to obtain a blue cone in the center of the flame. Later in this experiment (2. A), you will carry out a more complete analysis with the burner.

Weigh out separately about 0.1 g of sulfur and about 0.1 g of iron. Mix them thoroughly by grinding them together in a clean mortar. Did a reaction occur while you were mixing them? How could you prove this? (Answer Question 1B, 1–2.) Now put the mixture in a small test tube with your spatula and heat it strongly for 3 minutes at the tip of the blue cone, the hottest part of the flame. Let the tube cool and proceed as follows.

Your task at this point is to determine whether the material is still a mixture of iron and sulfur or whether a new compound was formed during the heating process. To the cooled material in the test tube add a few drops of dilute (6 M) hydrochloric acid. Observe whether a reaction occurs and carefully smell the gas which forms. Compare the odor (especially the intensity) with that from the second test in 1A. Now answer the questions. (Answer Question 1B, 3–7.)

Dispose of the entire test tube in the container provided; it cannot be cleaned.

2. Other Chemical and Physical Changes

In this part of the experiment you are to perform a number of short experiments and conclude, on the basis of some limited observations, whether a physical or a chemical change has occurred. Note that there are chart spaces and additional questions for each experiment.

A. Burning of Methane

Following the instructions in the demonstration in B, reignite your burner. The flame and heat in a Bunsen burner are produced when natural gas (methane, CH_4) combines with oxygen, O_2. On most burners the gas supply is controlled by a valve at the bottom of the burner, and the air supply is regulated by turning the barrel of the burner (figure 3.1).

Adjust the gas supply to obtain a flame about 7 or 8 cm high. Now decrease the air supply by turning the cylinder or sleeve until the flame becomes "cool" and luminous (yellow), and then increase it until a hot blue flame with a cone in the center develops. Repeat this procedure until you can quickly obtain either flame. (Answer Question 2a and Question 2A, 1–2.)

B. Heating of Camphor

Place about 0.2 g of camphor, $C_{10}H_{16}O$, in a test tube. Waft any vapors toward you. Note any observations. Now move the camphor in the tube to the cool side of the flame; hold the tube in this position until a change occurs. Immediately remove the tube and allow the tube to cool. Examine the physical properties of the material in the tube to determine what kind of change occurred. (Answer Question 2b and Question 2B, 1–4.)

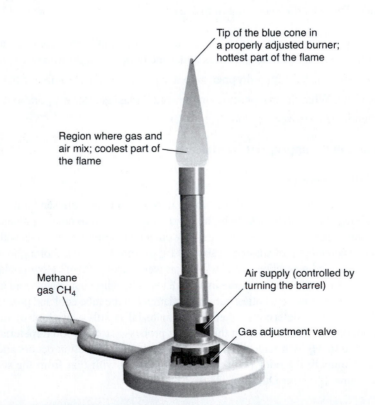

Tip of the blue cone in a properly adjusted burner; hottest part of the flame

Region where gas and air mix; coolest part of the flame

Methane gas CH_4

Air supply (controlled by turning the barrel)

Gas adjustment valve

▲ **FIGURE 3.1** A properly adjusted laboratory burner.

C. Iron and Copper Sulfate

Place a shiny iron nail in a test tube and add 1 mL of dilute (0.1 M) copper sulfate. After 15 minutes remove the nail and examine it for any evidence of chemical change. (Answer Question 2c and Question C, 1–2.)

D. Lead Acetate and Potassium Chromate

OPTIONAL: Your instructor may not wish to perform this experiment because of the toxicity of chromium.

   To a test tube containing 0.5 mL of 0.1 M potassium chromate, K_2CrO_4, solution add about the same volume of 0.1 M lead(II) acetate, $Pb(C_2H_3O_2)_2$, solution. Watch for the formation of chrome yellow, a pigment well known to paint formulators and artists. (Answer Question 2d and Question D, 1.)

Dispose of all waste from parts B, C, D in the appropriate containers.

E. Acid and Calcium Carbonate

Stir about 0.1 g of powdered calcium carbonate, $CaCO_3$, into 1 mL of water in a test tube. Note the properties of $CaCO_3$, particularly its solubility. Obtain about 1 mL of dilute (1 M) acetic acid, $HC_2H_3O_2$ (essentially identical to vinegar), and stir it slowly into the $CaCO_3$. Is a reaction occurring? What is the observable evidence? Mineral prospectors in the field use acid to detect the presence of calcite ore, which is primarily $CaCO_3$. (Answer Question 2e and Question E, 1–3.)

Dispose of waste by washing it down the drain.

F. Metabolism of Food

Most physiological reactions are too complex to be studied conveniently in the beginning laboratory. However, you can verify that foods such as table sugar produce carbon dioxide as they undergo metabolic reactions in the body. To do so, place 2 mL of a saturated solution of calcium hydroxide, $Ca(OH)_2$ (also called limewater), in a test tube and blow your breath into the solution through a clean plastic straw. Do this for a few minutes or until the solution becomes cloudy. A cloudy precipitate of calcium carbonate, $CaCO_3$, indicates the presence of carbon dioxide, CO_2, in your breath. What is the source of the carbon dioxide? (Answer Question 2f in the Observations and Results sheet and Question F, 1–2.)

QUESTIONS AND PROBLEMS

1.   Give three examples of exothermic reactions from everyday life.

2.   Describe what happens to the bonds in the following exothermic reaction. List which bonds are broken, and which are formed.

$$2H_2 + O_2 \rightarrow 2H_2O$$

Compare the relative strength of the bonds in the starting material to that of the bonds in the products.

3. Limestone is another form of $CaCO_3$. Sandstone is primarily silica or sand, SiO_2. On the basis of observations made in today's experiment, describe a test that would allow you to distinguish between limestone and sandstone.

4. Hard water deposits are largely $CaCO_3$. Suggest a way to remove these deposits from plumbing fixtures.

5. When you added hydrochloric acid to iron, what gas was formed? Give its name and formula and the number of atoms in each molecule.

6. The reaction between methane molecules and oxygen molecules (e.g., in the Bunsen burner) does not occur unless heat is supplied to start it. List other chemical changes (reactions) that require heat to initiate them.

7. The equation below, for the burning of methane, shows detailed bond structure.

$$
\begin{array}{c}
\text{H} \\
| \\
\text{H—C—H} \\
| \\
\text{H}
\end{array}
\quad + \quad
\begin{array}{c}
\text{O}=\text{O} \\
\text{O}=\text{O}
\end{array}
\quad \rightarrow \quad
\text{O}=\text{C}=\text{O}
\quad + \quad
\begin{array}{c}
\text{H—O—H} \\
\text{H—O—H}
\end{array}
$$

How many bonds were broken (count each dash or double dash as one bond)? How many bonds were formed?

EXPERIMENT

What Is a Chemical Reaction?

3

OBSERVATIONS AND RESULTS

1. Iron and Sulfur

A. *Properties of Iron and Sulfur*

1. List some physical properties for each element.

Iron   *Sulfur*

2. Was either element attracted to the magnet? If so, which?

3. Did either iron or sulfur react chemically with hydrochloric acid, HCl? If so, which?

4. What is the evidence for a reaction (i.e., formation of a new substance)?

5. Which element was affected by the carbon disulfide? Describe the change that occurred.

6. What was obtained after evaporation of the CS_2?

7. Was the change which occurred with CS_2 physical or chemical? What is your proof?

B. *Do Iron and Sulfur React?*

1. Did a reaction occur when the elements were mixed (prior to heating)?

2. Suggest two simple tests that you could use to establish whether a reaction occurred. (If you are uncertain whether a reaction occurred during mixing, use your tests to find out.)

3. What, if any, observations suggest that iron and sulfur react when they are heated?

4. Describe the odor of the gas which was formed when you added HCl to the contents of the test tube. The name of this gas is hydrogen sulfide.

5. Attempt to write a formula for hydrogen sulfide. Ask your instructor if your formula is correct.

   How does the fact that hydrogen sulfide is formed provide proof that a chemical reaction did occur when a mixture of iron and sulfur was heated? (Hint: How would a mixture of iron and sulfur have behaved toward acid?)

6. Complete the equation by adding a symbol.
$$Fe + \qquad \rightarrow FeS$$

7. Write an equation for the reaction of the material in the test tube with HCl.
$$\underset{\text{material in the test tube}}{} + 2HCl \quad \rightarrow \quad \underset{\text{odoriferous gas}}{} + FeCl_2$$

2. Other Chemical and Physical Changes

As you do each part of this experiment, indicate if there is a physical or chemical change and give the supporting evidence.

	Nature of Change	*Evidence*
a. Burning of methane		
b. Heating of camphor		
c. Iron and copper sulfate		
d. Lead acetate and potassium chromate		
e. Acid and calcium carbonate		
f. Metabolism of food		

A. *Burning of Methane*

1. The reaction that occurred in the Bunsen burner is written below. List the atoms that had bonds *broken* and those that had bonds *formed* between them.

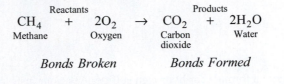

$$CH_4 + 2O_2 \rightarrow CO_2 + 2H_2O$$

Reactants: Methane, Oxygen   Products: Carbon dioxide, Water

Bonds Broken   *Bonds Formed*

2. Make a sketch of your burner. Indicate with arrows where each reactant enters the burner and where the products are formed.

B. *Heating of Camphor*

1. Describe the physical properties of camphor before heating.

2. What observable change was produced by heating?

3. Compare the properties of the cooled product with those of the original camphor.

4. What kind of change occurred? What is the evidence?

C. *Iron and Copper Sulfate*

   1. Observe and record the color of the *metal* (element) forming on the nail. There is only one common metal with that color.

   2. Now complete (with symbols) the following equation.

$$CuSO_4 \ + \ \underset{\text{Starting metal}}{\quad\quad} \ \rightarrow \ FeSO_4 \ + \ \underset{\text{New Metal}}{\quad\quad}$$

D. *Lead Acetate and Potassium Chromate*

   1. Complete the following equation, which describes the reaction leading to the formation of the yellow pigment.

$$\underset{\text{Lead acetate}}{Pb(C_2H_3O_2)_2} \ + \ \underset{\substack{\text{Potassium} \\ \text{chromate}}}{K_2CrO_4} \ \rightarrow \ \underset{\substack{\text{Potassium} \\ \text{acetate}}}{2\ KC_2H_3O_2} \ + \ \underset{\text{Lead chromate}}{\quad\quad}$$

E. *Acid and Calcium Carbonate*

   1. What is an important property of calcium carbonate, $CaCO_3$, with respect to its solubility in water?

2. What obvious physical property of acetic acid, $HC_2H_3O_2$, do you notice?

3. In considering whether a reaction occurred between $CaCO_3$ and $HC_2H_3O_2$, what evidence is there that new products have been formed?

4. Write an equation to describe the reaction between $CaCO_3$ and $HC_2H_3O_2$. You know the formulas for the reactants and for all of the products except the gas. Guess at the formula for the gas by examining your nearly completed equation.

$$\_\_\_ + \_\_\_ \rightarrow Ca(C_2H_3O_2)_2 + \underset{\text{Gas}}{\_\_\_} + H_2O$$

F. *Metabolism of Food*

1. What is the source of the carbon dioxide (refer to equation on page 28)?

2. Describe the test and explain the evidence which suggests that chemical changes involving foods were occurring in your body (refer to the equation on page 35).

The Calorie:
A Unit of Energy | 4

OBJECTIVES

1. To become familiar with the calorie as a unit of energy.
2. To learn how to measure heat effects in endothermic and exothermic processes.
3. To acquire knowledge of heats of transition as well as of ordinary heat requirements for effecting temperature changes.

DISCUSSION

Most of us become interested in calories for the first time when we decide to go on a diet. Yet few people who count calories for staying slim understand what a calorie is, or even that it is an energy unit. Does food possess energy? Does energy have anything to do with weight control? What are calories and how are they measured? You will be dealing with these questions in this experiment and later in Experiments 26 and 28, which are on Carbohydrates and Lipids.

Energy in the Body

Most foods are fuels for the body. Of course, certain components of our diet, such as vitamins, minerals, and some proteins, are used primarily for building and repairing tissues and for regulating functions rather than for supplying energy. The body requires a certain minimum daily amount of energy—around 1000 kilocalories (kcal)—just to stay alive. Energy is necessary for maintenance of body temperature and for mechanical performance such as heart and lung action and muscle motor activity. In fact, all body functions, including brain activity, are powered by energy. So we eat carbohydrates, along with some proteins and fats, to supply this needed energy.

Relatively inactive people use around 2000 kcal of energy per day, but a large laboring man may need as much as 5000 or 6000 kcal. Foods are rated according to the energy they supply when metabolized (burned) in the body.[1] These are the values used by calorie counters for diet control. The amount of energy released from foodstuffs, as listed in diets, is expressed in big Calories (note the capital *C*). One Calorie is equivalent to 1 kcal or 1000 cal. A bottle of diet soft drink containing 1 Cal really has 1000 cal.

No doubt you already understand that if you eat less food than you need for bodily functions, your body draws upon its inventory of fat (triglycerides) and, eventually, upon its protein to compensate for the deficiency. If you consume more food than you need, your body converts excess carbohydrates and proteins to fats, which are then stored in the tissues for future use. So overeating (and underexercising) results in a gain in body weight, and undereating results in a weight loss.

---

[1]Generally, foods are capable of being burned as fuel in an ordinary flame. The pathway is different, but the overall chemical reaction is the same as that which occurs in the body, with identical amounts of energy being liberated. The combustion reaction (exothermic) for ordinary sugar in the laboratory (and in the body) is

$$C_{12}H_{22}O_{11} + 12\,O_2 \rightarrow 12\,CO_2 + 11\,H_2O + \text{energy}$$

| 342 g | 384 g | 528 g | 198 g | 1340 kcal |

Stored fat in the body is not undesirable unless it is excessive. The right amount is not only esthetically pleasing but has value as insulation and padding. It may even be reassuring to know that if our food supply were to be abruptly cut off, most of us could continue to function for at least a month by living off our fat inventory.

Basic Concepts of Energy

The study of energy and its transformations, called thermodynamics, is a vast and important area of knowledge. Energy, the capacity to do work, is expressed in several different units, such as ergs, joules (J), foot-pounds, British thermal units (BTUs), and calories. The calorie is one of the favorite units for expressing heat energy, partly because it is readily understood and easily defined. A calorie is the amount of heat required to raise the temperature of 1 g of water 1°C. Since this amount depends slightly upon the temperature at which the measurement is made, a precise definition requires that the temperature range be specified. Thus, a calorie is defined as the amount of heat needed to raise 1 g of water from 14.5 to 15.5 °C. Many scientists prefer to use the joule as the unit of energy because it is strictly a metric unit (4.18 joules = 1 calorie). The *specific heat* of a substance is the quantity of heat energy required to raise 1 g of that substance 1°C. From the definition of a calorie, it is obvious that the specific heat of water is 1 calorie per gram per degree (cal/g-deg). Each substance has its own unique specific heat.

Forms of Energy

Energy is manifested in many forms, including heat, light, mechanical energy, electrical energy, and chemical energy. Energy may be converted from one form to another, but it is not created or destroyed. Food has energy stored as chemical energy. The metabolic process (metabolism) converts part of this chemical energy to heat and mechanical energy. When a flashlight is turned on, chemical energy in the cells (battery) is converted to light energy. Electrical energy is easily converted to heat, light, or mechanical energy. When we say energy is consumed, we mean that energy in a highly useful form, such as chemical or electrical energy, is converted to a less useful kind, such as low-temperature heat. Ultimately, heat is lost from the earth by being radiated to outer space.

Heat is the most familiar form of energy. All things contain some heat since all matter is at a temperature higher than absolute zero.[2] Heat always flows from regions of higher temperature to regions of lower temperature. Cold, which is nothing but the absence of heat, cannot flow. If an object feels cold to the skin, heat is flowing from the skin to the object.

Measuring Calories

You can measure the heat effect for a process, exothermic (heat evolved) or endothermic (heat absorbed), by staging the heat transfer in a reservoir of water. The measurement is based on the fact that the specific heat (the amount of heat required to raise 1 g of a substance 1°C) of water is 1.00 cal/g-deg.

Heat liberated in an exothermic process flows into the reservoir of water, thus raising its temperature. In an endothermic process heat is absorbed from the water, lowering its temperature. The total heat energy released into or absorbed from the reservoir can be calculated from the temperature change and the total mass of water present, according to the following relationship.

Calories of heat = mass of water (g) × temperature change (°C) × specific heat (cal/g-deg)

In this experiment you will study the heat effects associated with such processes as heating water, dissolving a salt, melting ice, and condensing steam. In considering all heat (or other energy) transfers, you should keep in mind one important principle: The heat energy which leaves one object is the amount gained by the other.

---

[2]At absolute zero (−273.16°C) —which is only approached, never reached—there is no more heat to be extracted.

Energy is never lost; it merely changes form or location according to the law of conservation of energy.[3] For example, the heat gained by water when it is heated by a Bunsen burner is given up by the hot gases in the flame. When ice is placed in water, the heat lost by the water as it cools is equal to that which is gained by the ice as it melts. In an exothermic chemical reaction, the heat absorbed by the water is equal to the chemical energy released by the reactants. If a process is reversed, the heat effects are numerically the same. If 50 cal are required to warm a sample to a certain temperature, then 50 cal will be released when the sample is returned to its original temperature.

Heats of Fusion and Vaporization

A relatively large amount of energy is required just to convert a substance from a solid to a liquid (fusion). This quantity of heat is called the *heat of fusion*.[4] For water, the temperature of both ice and liquid is 0°C at the freezing point (melting point). The heat of fusion for water is 80 cal/g. If more heat is supplied than is required to melt the ice, the temperature of the resulting liquid will rise above 0°C.

We can make a similar statement about the conversion of liquid water to steam. Exactly 540 cal are required to vaporize 1 g of water at 100°C. This is the heat of vaporization of water. Heats of fusion and of vaporization are specific properties of a given substance.

If you eat an ice cube, sufficient heat is supplied by your body to melt the ice. Your body may cool momentarily, but metabolized food immediately furnishes enough heat to restore normal body temperature. The amount of heat required to melt ice in the body is the same as that required to melt ice under laboratory conditions.

In this experiment you will determine the heat of fusion of ice by mixing a weighed amount of ice with a known amount of liquid water and noting the temperature change of the liquid. The heat released by the liquid as it cools (1 cal/g-deg) is sufficient to melt a certain quantity of ice. The amount of ice that can be melted depends on its heat of fusion. The following relationship holds:

Heat supplied by water = heat absorbed by ice

Or

Mass of water cooled (g) × temperature drop (°C) × specific heat (cal/g-deg)

= mass of ice melted (g) × heat of fusion (cal/g)

----

[3]The first law of thermodynamics states that in all chemical transformations energy is neither created nor destroyed.

[4]The heat of fusion is released when the liquid is converted to a solid. There is no change in temperature during the fusion process.

ENVIRONMENT, CULTURE, AND CHEMISTRY
Future Energy Sources

Few subjects are as important or as contentious as that of acquiring and regulating our energy sources. People with an understanding of chemistry will surely be required to assist in solving this huge problem and to help make the necessary political judgments. For starters, here are three important considerations: First, metallic ores and fossil fuel deposits (petroleum, coal, natural gas) have been removed from the earth at a prodigious rate. Recycling of metals and materials has begun, and in theory, and with sufficient commitment and substantial payment, 100 percent recycling can be approached. So at least in theory, the supply of metals need not be depleted. However, fossil fuels cannot be recycled. When they are gone, they are simply gone; there is no way of replacing them, since the conditions on earth which produced fossil fuels from living materials are not likely to be repeated, at least for millions of years. Nuclear energy sources (uranium and other radioactive materials) are also non-recyclable, although there appears to be a plentiful supply for a long time. Second, this leaves renewable energy sources as the alternative for the short term and even the long term. In spite of the optimism of politicians, all reasonable calculations project that alternative energy will come at a higher cost, even with the prospect of significant innovation through research and development. Solar energy, though plentiful, is anything but cheap. The problem and expense comes from building and maintaining the devices to harness that energy from the heat of the sun. Furthermore, a huge area of terrain will need to be covered. Energy from wind, again, has all of the problems of the cost of devices and the clutter of the landscape, and will only supply a small part of our energy needs. Someday, as conventional energy sources (fossil fuels) become depleted, and consequently more expensive, these alternative sources will become more competitive, and of necessity, we will also begin to seriously conserve energy. Third, there is one great hope for continued cheap abundant energy, controlled nuclear fusion, where small nuclei like hydrogen fuse together to give another atom (in this case, helium) and energy. Since this is theoretically possible (it is the source of the sun's energy), massive research has occurred in this area for the past sixty years, but with no breakthrough in sight. Nuclear fusion would be clean, radiation-free energy, producing no greenhouse gases. The stakes are huge.

Pre-Laboratory Questions | 4

1. Make a list of as many units of energy as you can.

2. Define *calorie*, *specific heat*, and *heat of fusion*.

3. Explain what happens to the heat energy as it is supplied to a pot of boiling water at 100°C. What is heat of vaporization?

4. How many calories of heat must be added to raise the temperature of 40 g of water from 25 to 50°C?

5. What is the final temperature of 100 g of water, initially at 20°C, to which 1000 calories (cal) are added?

EXPERIMENT

1. The Absorption of Heat by Water—Heat Storage and Release (A Calculation Exercise Only)

To introduce you to heat calculations in the laboratory let us simulate on paper the experience of a student who wanted to put some heat energy in a hot water bottle for a sick friend whose feet were "freezing" in a cold bedroom. The student put 800 mL of ordinary tap water at 25°C in a beaker on a hot plate. When the temperature had risen to 85°C the water was poured into the hot water bottle for the friend. How much heat was added to the water by the hot plate (1e)? If the hot water bottle remained in the bed at the friend's feet until the temperature had dropped to 35°C, how much heat had been absorbed by the friend's feet and bedding? Neglect the heat absorbed by the bottle itself (which you could not ignore when doing very precise work). (Record your data in 1a, b, c, f, g and calculate answers for 1d, e, h, i on the Observations and Results sheet.)

2. Heat of Solution of a Salt

Many physical processes as well as chemical reactions are exothermic or endothermic. When salts dissolve in water there is a breakdown of the crystal structure and an interaction (hydration) of water molecules with ions of the salt. In this part of the experiment you will measure the heat effect (a physical quantity) associated with the solution of a particular salt in water. Determine the heat of the solution of either calcium chloride, $CaCl_2$, or ammonium chloride, NH_4Cl, as described in the next paragraph, while your neighbor does the other. Then compare results and indicate in each case whether the process is exothermic or endothermic. (Option: If time permits, do both parts yourself.)

Weigh out accurately about 4 g of your salt and record the weight. Accurately measure 30 to 40 mL of water into a small styrofoam cup (a 4 oz. size works well). (A styrofoam cup is a good heat insulator, and it prevents the energy involved in this process from escaping to the surroundings.) Determine and record the initial temperature of the water, reading the thermometer to a fraction of a degree. Dissolve the salt in the water. Mix thoroughly, and then record the final temperature of the solution. Note the temperature change. Calculate the calories liberated or absorbed per gram of salt. Use the total weight of solution in determining the number of calories. Assume that the specific heat of the solution is the same as that of water. (Record your data in 2a, b, d, e and answer Questions 2c, f, g, h, i.)

Dispose of waste in the sink with an ample amount of water.

3. Heat of Fusion of Ice

Half-fill a 250-mL beaker with tap water and heat it to about 60°C. Use a portion of the heated water to warm a graduated cylinder, and then pour 25 to 30 mL of the water into the cylinder and insert a thermometer. Fill a styrofoam cup with at least 50 g of chipped ice and pour off any liquid present. Immediately read and record the volume and temperature of the warm water and pour the measured water over the ice. Stir vigorously until the water temperature reaches 2°C or less. Record this final temperature and immediately pour all of the ice water back into the graduated cylinder, draining the ice well (figure 4.1). Record the new volume of water. The increase in liquid volume came from the ice. The heat necessary to melt the ice was provided by the warm water. Calculate both the amount of ice melted and the heat required to melt the ice. From this, calculate the amount of heat required to melt each gram of ice, the heat of fusion. If time permits, part 3 should be repeated. (Record data in 3a, b, c, e and calculate answers for Questions 3d, f, g, h, i.)

4. Heat Transfer in the Body: Illustrations and Calculations

The following simple operations illustrate some heat transfer processes which can occur in the body. From what you already have learned, calculate the amount of heat supplied (or absorbed) by the body in each case.

a. Get a piece of ice about a gram in mass. Rub it on your arm until it melts. Do you feel the heat leaving your body to melt the ice? How much heat was required to melt the 1 g sample of ice and warm the resulting water to body temperature (37°C or 98.6°F). (Answer Question 4a. Calculate this as a two-step process. Melt the ice to water at 0°C, then warm the water to 37°C.)

b. Spread about 1 g (1 mL) of tap water over your hands and arms and allow them to dry by evaporation. Calculate the amount of heat supplied by your body to warm the water from room temperature to body temperature and to evaporate it, assuming you used exactly 1 g of water. (Assume the heat of vaporization of water is 540 cal/g at body temperature.) (Answer Question 4b. Again do a two-step calculation.)

c. Rubbing alcohol (isopropyl alcohol) is sometimes applied to the skin to help reduce a fever (remove heat) by evaporation of the alcohol. Repeat part 4b, using room-temperature isopropyl alcohol in place of water (assume the heat of vaporization of isopropyl alcohol at body temperature is 173 cal/g). Neglect any difference in the specific heats of the two liquids. (Answer Question 4c. Again a two-step calculation.)

d. You may have had a cup of hot coffee at breakfast this morning. Assume you drank 240 mL of coffee (hot water) at 80°C and that this amount was cooled by your body to the normal 37°C. How many calories were supplied to your body? (Answer Question 4d.)

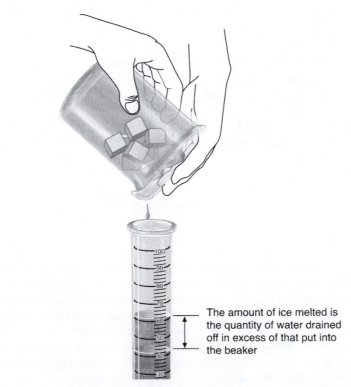

The amount of ice melted is the quantity of water drained off in excess of that put into the beaker

▲ **FIGURE 4.1** Draining and measuring the volume of ice water.

QUESTIONS AND PROBLEMS

1. Suppose the sun were used as a heat source to warm some water in a solar-heated house or swimming pool. How much solar energy (in kilocalories) would be stored in a 500-L tank whose temperature was raised from 30 to 70°C?

2. How many calories would be needed to convert 5 g of ice at 0°C to steam at 100°C?

3. What would be the final temperature of the water resulting from the addition of 20 g of steam at 100°C to 80 g of ice at 0°C?

EXPERIMENT

The Calorie: | 4
A Unit of Energy

OBSERVATIONS AND RESULTS

1. The Absorption of Heat by Water: Heat Storage

a. Milliliters (grams) of water in beaker _____ mL (g)

b. Final temperature of heated water _____ °C

c. Initial temperature of water _____ °C

d. Temperature increase (1b – 1c) _____ °C

e. Number of calories absorbed by the water $(1a \times 1d \times 1)$ _____ cal

f. Number of kilocalories absorbed _____ kcal

g. Final temperature of water after use in bed _____ °C

h. Temperature decrease (1b – 1g) _____ °C

i. Number of calories released by water in bed $(1a \times 1h \times 1)$ _____ cal

2. Heat of Solution of a Salt $CaCl_2$ NH_4Cl

a. Grams of salt _____ g _____ g

b. Milliliters (grams) of water in cup _____ mL(g) _____ mL(g)

c. Grams of solution (2a + 2b) _____ g _____ g

d. Final temperature of solution _____ °C _____ °C

e. Initial temperature of water _____ °C _____ °C

f. Temperature change _____ °C _____ °C

g. Total calories of heat energy evolved or
   absorbed (show your calculations) _____ cal _____ cal

h. Was the process exothermic or
   endothermic? _____ _____

i. Heat of solution per gram of salt (2g/2a);
   indicate whether absorbed or evolved _____ cal/g _____ cal/g

3. Heat of Fusion of Ice

	Trial 1		Trial 2	
a. Milliliters (grams) of warm water used	_____	mL(g)	_____	mL(g)
b. Initial temperature (warm water)	_____	°C	_____	°C
c. Final temperature (ice water)	_____	°C	_____	°C
d. Temperature change (3b − 3c)	_____	°C	_____	°C
e. Milliliters (grams) of ice water drained from ice	_____	mL(g)	_____	mL(g)
f. Grams of ice melted (3e − 3a)	_____	g	_____	g
g. Calories required to melt ice (3a×3d)	_____	cal	_____	cal
h. Calories required to melt 1 g of ice (3g/3f) (heat of fusion of water)	_____	cal	_____	cal
i. Difference between your experimental heat of fusion and the actual value	_____	cal/g	_____	cal/g

Suppose the quantity of ice you used had been eaten. Would one slice of bread, capable of supplying 60 kcal of heat, be sufficient to melt the ice? How much ice could it melt?

4. Heat of Transfer in the Body

a. Heat supplied by the body to melt 1 g of ice and warm the water to body temperature	_____	cal
b. Heat supplied by the body to warm 1 g of water to body temperature and to vaporize it	_____	cal
c. Heat supplied by the body to warm 1 g of isopropyl alcohol to body temperature and vaporize it	_____	cal
d. Heat supplied to the body from 240 mL of coffee at 80°C	_____	cal

Reliability of Data: The Composition of a Compound | 5

OBJECTIVES

1. To develop a basis for treating experimental data and evaluating its validity.
2. To learn some quantitative laboratory techniques.
3. To demonstrate that a compound has a definite composition by weight.
4. To become familiar with a class of compounds called hydrates.

DISCUSSION

In professional and everyday life, important decisions are based too frequently on careless and erroneous interpretations of data. Unfortunately, many influential decision makers have not learned about the valid treatment of data. One of the purposes of the first experiment in this manual was to help you develop a sense of reliance upon the results of your laboratory measurements. One of the purposes of this experiment is to help you realize that there should be limits to that confidence. Since every measurement, no matter how carefully made, inherently contains some error, you should learn to scrutinize experimental and statistical data. After thorough evaluation of data, you should be prepared to accept the results up to their limits of confidence. The ultimate goal is to obtain data so accurate that the error is minimal.

Hydrates

In today's experiment, you are going to determine the percentage of water in copper(II) sulfate pentahydrate, $CuSO_4 \cdot 5H_2O$. This compound, also known as blue vitriol, is representative of a class of compounds called *hydrates*. These are salts which contain a definite amount of water in the salt crystals; the amount is specified in the formula of the hydrate. Examples of hydrates[1] are barium chloride dihydrate, $BaCl_2 \cdot 2H_2O$, and copper(II) sulfate pentahydrate, $CuSO_4 \cdot 5H_2O$. The water molecule, called a ligand, is usually bonded directly to the metal cation by a covalent bond (often relatively weak), as in the following example for $Ba(H_2O)_2{}^{2+}$ ion.

$$\left[\begin{matrix} H \\ H \end{matrix} \!\!>\!\! O - Ba - O \!\!<\!\! \begin{matrix} H \\ H \end{matrix} \right]^{2+}$$

---

[1] The dot in a hydrate formula indicates that a definite number of water molecules is part of the formula for the compound.

The complex positive ion then has normal ionic bonding with the negative chloride ions. Occasionally, the water may be bonded to the anion. In any case, the ligand bond is not exceedingly strong, and when hydrates are heated, water is lost, leaving the anhydrous (water-free) salt.

Composition of a Compound

In a compound the elements are present in definite ratios and in definite percentages by weight. This is known as the law of constant composition (also called the law of definite proportions). You will use the decomposition of a hydrate (a pure compound) to simpler substances to demonstrate this law. In decomposition, the weaker bonds between atoms are broken in each molecule (formula unit) of the sample. For a particular decomposition, all reactant molecules produce identical sets of product molecules. If one of the products of the reaction is found to represent 30% by weight of the original starting material, then that means that the decomposition of each reactant molecule yields a piece which represents 30% by weight of the original molecule. Whether one decomposes a single molecule, 100 molecules, or billions of molecules, 30% of the weight of the starting material shows up in the form of this particular product.

In today's experiment you will heat a weighed quantity of the blue $CuSO_4 \cdot 5H_2O$ crystals until the water in the hydrate is driven off as steam, leaving as a residue the white anhydrous salt. This transformation (reaction) may be represented by the following equation (Δ is the symbol for heat).

$$CuSO_4 \cdot 5H_2O \xrightarrow{\Delta} CuSO_4 + 5H_2O$$
$$\text{Blue solid} \qquad\qquad \text{White solid} \quad \text{Vapor}$$

The difference in weight between the hydrate and the anhydrous salt is equal to the weight of water in the hydrate: It is the amount of water which was vaporized in the decomposition. The percentage of water is easily calculated from these weights.

$$\text{Experimental percentage of water} = \frac{\text{weight of water lost on heating}}{\text{weight of original sample of } CuSO_4 \cdot 5H_2O} \times 100$$

Accuracy and Precision of Data

Two words describe the quality of experimental data: accuracy and precision. *Accuracy* measures the correspondence of the experimental value with the true value. The more nearly the experimental value corresponds to the true value, the more accurate it is. For example, if a sample is known to contain 15.5 g of iron, then an experimentally determined value of 15.7 g is more accurate than a value of 15.0 g.

Accuracy is usually expressed as *percentage error*, which decreases as the accuracy increases.

$$\text{Percentage error} = \frac{\text{difference between correct value and experimental value}}{\text{correct value}} \times 100$$

For our example above, the percentage error for the first experimental value is

$$\frac{0.2}{15.5} \times 100 = 1\% \text{ (Rounded to agree with the significant figures of the numerator)}$$

For the second experimental value, the percentage error is

$$\frac{0.5}{15.5} \times 100 = 3\%$$

Precision indicates how well measurements in a series of experimental values agree.

Experimental Values of Series A	*Experimental Values of Series B*
15.4	15.2
15.7	16.1
15.4	14.8
15.5	15.9

In series A, the values are grouped more closely together than are the values of series B. The *range* for the values in series A is 15.4 (lowest value) to 15.7 (highest value). For series B, the range is 14.8 to 16.1. Series A is more precise than series B.

In evaluating precision, we are considering how much each value in a series deviates or differs from the mean (or simple average) of all the values of the series.

$$\text{Mean value of series A} = \frac{15.4 + 15.7 + 15.4 + 15.5}{4} = 15.5$$

$$\text{Mean value of series B} = \frac{15.2 + 16.1 + 14.8 + 15.9}{4} = 15.5$$

For both series, the mean value is 15.5. For each value of each series we can now calculate the deviation from the mean value. (Note that whether the deviation is negative or positive is not considered.)

Deviations from the Mean for Series A	*Deviations from the Mean for Series B*
$15.5 - 15.4 = 0.1$	$15.5 - 15.2 = 0.3$
$15.7 - 15.5 = 0.2$	$16.1 - 15.5 = 0.6$
$15.5 - 15.4 = 0.1$	$15.5 - 14.8 = 0.7$
$15.5 - 15.5 = 0.0$	$15.9 - 15.5 = 0.4$

For each series, we can also calculate the average deviation.

$$\text{Average deviation for series A} = \frac{0.1 + 0.2 + 0.1 + 0.0}{4} = 0.1$$

$$\text{Average deviation for series B} = \frac{0.3 + 0.6 + 0.7 + 0.4}{4} = 0.5$$

A low average deviation denotes a high degree of precision and occurs when a large number of values fall within a narrow range. Frequently, the average deviation is expressed as a percentage of the mean value. For series A,

$$\frac{0.1}{15.5} \times 100 = 0.6\% \qquad \text{For series B,} \qquad \frac{0.5}{15.5} \times 100 = 3\%$$

Generally, high precision (low average deviation) is indicative of high accuracy, although exceptions arise where a constant procedural error is repeated.

Often one value in a group of results differs significantly from the others. In such cases, it is likely that some inadvertent error was made in weighing, recording of data, and so forth. Should this value be discarded? Here is a simple rule that you may use in deciding whether a value should be discarded: *If the deviation of the suspected value from the mean is equal to three times (or more) the average deviation from the mean of the other (three or more) values, the suspected value should be discarded.*

An example will help to explain the test. A sample of barium chloride dihydrate, $BaCl_2 \cdot 2H_2O$, was heated to remove the water. The following are experimental results for the percentage of water in the compound: 14.70%, 14.76%, 14.68%, 14.52%, and 14.63%.

1. Which value is suspect? The 14.52 value appears to deviate significantly.

2. What is the mean of the other four values?

$$\frac{14.70 + 14.76 + 14.68 + 14.63}{4} = 14.69$$

3.   What are the deviations of these four values from the mean?

14.70	14.76	14.69	14.69
−14.69	−14.69	−14.68	−14.63
0.01	0.07	0.01	0.06

4.   What is the average of the four deviations?

$$\frac{0.01+0.01+0.07+0.06}{4}=0.04$$

5.   How much does the suspected value deviate from the mean of the others?

$$14.69-14.52=0.17$$

6.   Is the deviation of the suspected value (step 5) equal to three times (or more) the average deviation of the other values (step 4)? The value of 0.17 is greater than 0.12 (0.04×3).

7.   Can the suspected value be discarded? Yes. The mean of the four remaining values is 14.69 (step 2). If the true value for the percentage of water in the hydrate is 14.72%, then the percentage error can be calculated.

$$\text{Percentage error}=\frac{14.72-14.69}{14.72}\times100=0.20\%$$

You must realize that precise work requires many refinements in technique, elaborate equipment, and much time and expense. Today's experiment is limited by what can be done, simply, in one laboratory period. Some applications warrant the cost of obtaining very precise data. In other situations only a rough estimate is needed. Learn to recognize that different situations have different requirements for accuracy and precision.

ENVIRONMENT, CULTURE, AND CHEMISTRY
Reliability of Environmental Data

In protecting ourselves, our health, and our environment, it is essential to rely on the best possible data. The more confidence that can be placed in the data, the greater the certainty that action is warranted and vice versa. Consider pharmaceutical drugs: Extensive trials are first made with test animals and then on humans to insure the highest possible level of safety. Occasionally, however, after perhaps millions of uses by consumers, evidence emerges that the safety is not in line with what was once assumed (Avandia, the anti-diabetes drug, and the quandary over estrogen therapy come to mind). Then a reevaluation is required to weigh the benefits against the new risk factor, considering the seriousness of the disease, and also taking into account any alternatives available—remember, in the real world, tradeoffs are often messy but a necessary part of life. On the other extreme, it is sometimes necessary to draw conclusions based on only a few or no experiences or experiments. For example, only a few airplane crashes, bridge collapses, nuclear explosions, and space ship failures occur, or can be studied effectively, to guide our safety regulations. Since there is only one earth on which to observe climate factors, none on which to scientifically vary climate factors, and none on which to conduct experiments to establish any theories, our knowledge will of necessity be limited. This is why, even among scientists trained in objectivity, there are vast differences of opinion on such important topics as global warming and climate change. In the last analysis, all that finally stands is informed opinion taking into account the best of both theories and observations.

Pre-Laboratory Questions | 5

1. Distinguish between a mixture and a compound. Give three examples of each.

2. A student reported 15.12% water in $BaCl_2 \cdot 2H_2O$. The true value is 14.72%. What is the percentage error?

3. Suppose four weighings of a crucible were found to be 10.231 g, 10.278 g, 10.256 g, and 10.363 g. Obtain the average (mean) weight of the crucible. Is the 10.363-g value as reliable as the other three weighings? Explain.

4. For each formula unit of sodium sulfate decahydrate, $Na_2SO_4 \cdot 10H_2O$, (a) How many atoms of each kind are present? (b) What percentage of the atoms are oxygen atoms? (c) How many water molecules would be liberated for every molecule of the hydrate that is decomposed by heating?

5. Describe how you would make 2.35 g of a mixture of sand, clay, and ash in the following percentages, respectively: 24.64%, 4.23%, and 71.13%.

EXPERIMENT

1. Place a clean crucible, without a cover, on a clay triangle which has been adjusted to the proper height above a Bunsen burner. (See figure 5.1.) Heat the crucible for about 5 minutes with a low flame to remove any volatile or flammable material. Transfer the crucible, using crucible tongs, to a wire gauze and allow it to stand until it is cool to the touch. Carry the crucible to the balance with tongs, weigh it carefully, and record the weight on line 1b. Put about 1 g of copper(II) sulfate pentahydrate, $CuSO_4 \cdot 5H_2O$, directly into the crucible. Accurately weigh the crucible and its contents and record the weight on line 1a on the Observations and Results sheet. Transfer the crucible and the hydrate to the clay triangle and heat with a low flame. Observe how the blue color changes to off-white. When the color change is complete (about 5 minutes), cool and weigh the crucible and contents (anhydrous copper(II) sulfate). Record the weight on line 1.d. Calculate the percentage of water in the hydrate. (Calculate answers for line 1c, e, f.)

    Compare your experimental result with the theoretical percentage of water in the hydrate, which can be calculated from the formula $CuSO_4 \cdot 5H_2O$. The atomic weights of the elements are copper, 63.5; sulfur, 32.0; oxygen, 16.0; and hydrogen, 1.0. The total atomic weights for the elements in one molecule of copper(II) sulfate pentahydrate are

▲ **FIGURE 5.1** A crucible positioned on a clay triangle at the proper height for heating with a laboratory burner.

$$\text{Total atomic weight of copper} \quad = 63.5 \times \ 1 = \ 63.5$$
$$\text{Total atomic weight of sulfur} \quad = 32.0 \times \ 1 = \ 32.0$$
$$\text{Total atomic weight of oxygen} \quad = 16.0 \times \ 9 = 144.0$$
$$\text{Total atomic weight of hydrogen} \quad = \ 1.0 \times 10 = \ 10.0$$
$$\text{Total molecular weight} = 249.5.$$

Thus, the molecular weight of copper(II) sulfate pentahydrate is 249.5. The theoretical percentage of water can be found as follows;

$$\text{Percentage of } H_2O = \frac{\text{molecular weight of } 5H_2O}{\text{molecular weight of } CuSO_4 \cdot 5H_2O} \times 100$$

$$= \frac{5(2+16)}{63.5+32+144+10} \times 100$$

$$= \frac{90}{249.5} \times 100$$

$$= 36.0\%$$

(Calculate answers for 1g, h, i and answer question j.)

2. Remove the crucible cover and note the color of the anhydrous salt. Add a few drops of water to the anhydrous salt. Does the blue color return? If so, what is responsible for it? Note any additional observations. (Answer Question 2, a–c.)

Dispose of waste in the appropriate container.

3. Collect results from other students in part 3a and analyze the results in part 4a–e on the Observations and Results sheet.

QUESTIONS AND PROBLEMS

1. Calculate the theoretical percentage of water in gypsum (calcium sulfate dihydrate, $CaSO_4 \cdot 2H_2O$).

2. A student reported the following values for the percentage of copper in $CuSO_4 \cdot 5H_2O$: 24.40, 24.32, 24.45, 24.45, and 24.27%. What is the range? the mean? the average deviation?

3. Calculate the theoretical (true) percentage of copper in question 2 and determine the percentage error.

4. A student obtained the following results for the percentage of water in aluminum sulfate octadecahydrate, $Al_2(SO_4)_3 \cdot 18H_2O$: 48.70, 48.65, 49.50, and 47.99%. Determine the experimental error (%) in these data. Make use of the discard test and show all calculations.

5. Recall one or more situations in the past few years where the accumulation of experimental data eventually led to the removal of a consumer product from the market. Were the data valid?

EXPERIMENT

Reliability of Data: The Composition of a Compound

5

OBSERVATIONS AND RESULTS

1. a. Weight of crucible and hydrate _____ g

   b. Weight of crucible _____ g

   c. Weight of hydrate (a − b) _____ g

   d. Weight of anhydrous $CuSO_4$ and crucible _____ g

   e. Weight of anhydrous $CuSO_4$ (d − b) _____ g

   f. Weight of water in hydrate (c − e) _____ g

   g. Using your experimental data, calculate the percentage of water in copper(II) sulfate pentahydrate (show all calculations).

   h. From the formula of $CuSO_4 \cdot 5H_2O$, what is its theoretical percent H_2O?

   i. Calculate the percentage error in your determination, comparing the experimental and the theoretical values for the percentages of water in the hydrate (show all calculations).

   j. List as many possible sources of error for this experiment as you can. Tell how each of these would affect the experimental value for the percentage of water in the hydrate. What do you think are the major sources of error in your experiment?

2. a. What is the color of the anhydrous salt?

b. Does the color return when water is added to the anhydrous salt? What is responsible for the blue color?

c. Write an equation for the reaction that occurs when water is added to the anhydrous salt.

3. a. Record the percentages of water from your determination and the determinations of at least five other students (indicate the name of the student beside the data).

Student's Name	*Percentage of Water*
_____	_____
_____	_____
_____	_____
_____	_____
_____	_____
_____	_____

4. Carry out the following analysis on the precision and accuracy of these data.

a. Determine the range (lowest to highest percentages).

b. Use the "discard test" to determine if one (or more) of the values can be discarded (show all calculations).

c. Determine the mean, after discarding any suspect values, and use this mean, which is the experimental value for the percentage of water in the hydrate, to determine the percentage error, taking 36.0% water as the true value.

Mean _____ %

Percentage error _____ %

d. One would expect the accuracy of data to improve with an increase in the number of determinations. Was the percentage error for the group study less than for your own determination? If you did not see an improvement with the group data, give an explanation for this.

e. How do the data from this experiment support and illustrate the law of constant composition?

Getting Acquainted with Metals | 6

OBJECTIVES

1. To become familiar with the properties of various metals, and especially with the differences in their reactivities.
2. To learn to relate the activity of a metal to its position in the periodic chart.
3. To understand clearly the distinction between free metals and metals in the combined state.
4. To appreciate the benefits and hazards of metals.

DISCUSSION

About three-fourths of the 118 elements are metals. Of these, at least 20 are fairly common and are of major importance. To get acquainted with some properties of various metals, peruse table 6.1. You will study some of the chemical properties of metals in today's experiment. Metals in their elemental form are commonly described as lustrous. They are good conductors of heat and electricity and are capable of being deformed without breaking (malleable). The fusing (melting together) of particular metals gives alloys which have a wide variety of properties and uses. Steels are alloys of iron with other metals and carbon. Stainless steels contain nickel and chromium. Brass contains mostly copper and zinc, whereas bronze is primarily copper and tin.

Metals and Their Compounds

Metals and nonmetals combine in many ways to make compounds, some of which are soluble in water. In most compounds the metal exists in the form of positive ions and the nonmetal as negative ions.[1] Examples, showing the charges on the ions, are sodium chloride, Na^+Cl^-, and copper(II) sulfate, $Cu^{2+}SO_4^{2-}$. Keep in mind the distinction between the free metal (the element) and the combined form of the metal (the ion); the former consists of neutral metal atoms and the latter pairs with negatively charged ions to form compounds. There is a vast difference between copper metal (red, lustrous, solid, insoluble in water, somewhat soft and flexible) and copper ions in copper sulfate (blue crystal, brittle, soluble in water)[2] (figure 6.1). When you use copper(II) sulfate solution in this experiment, note how copper (as ions) in solution is very unlike copper (as atoms) in metallic form. This experiment focuses on the properties of free metals. A later experiment (experiment 16) deals with reactions of some metal ions in solution.

---

[1]Generally those compounds of the metals whose negative ions are oxides, O^{2-}, or hydroxides, OH^-, are called bases. The others are called salts.

[2]The properties of a salt are due to both kinds of ions, but some of the properties can be ascribed to particular ions. For example, the blue color of copper(II) sulfate solutions is due to the copper(II) ions; sulfate ions are colorless.

TABLE 6.1 Some Metals and Their Properties

Metal	Symbol	Property	Uses
Iron	Fe	Strong, tough, corrodes	Main ingredient in steel
Nickel	Ni	Resists corrosion	Coinage, alloy for stainless steel
Chromium	Cr	Shiny, resists corrosion	Chrome plating
Zinc	Zn	Forms protective coating of ZnO	Galvanizing coating
Radium	Ra	Radioactive	Cancer treatment
Aluminum	Al	Light and strong	Airplanes, window frames
Uranium	U	Fissionable	Energy source
Magnesium	Mg	Light and strong	Auto wheels, luggage
Mercury	Hg	Dense liquid at room temperature	Thermometers, barometers
Sodium	Na	Very active, soft	Heat transfer medium
Copper	Cu	Red color, good electrical conductor	Electrical wiring
Silver	Ag	Excellent electrical conductor	Electrical contacts, mirrors, jewelry
Gold	Au	Yellow metal, soft	Coinage, instruments, dentures
Platinum	Pt	Inert, high melting	Jewelry, instruments
Tantalum	Ta	Resists attack by acids	Synthetic skulls and bone parts
Tungsten	W	Very high melting	Lightbulb filament
Tin	Sn	Resists corrosion	Coating for steel cans
Lead	Pb	Low melting, dense, soft	Plumbing, fishing weights

In nature metals usually are found in combination with oxygen or other nonmetals. Only a few of the less reactive ones, such as gold and sometimes silver and copper, are found in the free metallic (native) state.

Activity of Metals

One of the most important characteristics of a metal is its activity (reactivity, or ability to react to form compounds). Metals range widely in activity, from vigorously reactive cesium, potassium, and sodium, to quite inactive (inert) platinum, gold, and silver. Since the latter resist oxidation, they are often called the noble metals. The coinage metals include gold and silver along with some metals of lesser value, such as copper and nickel. Many metals become oxidized by reacting with oxygen in the air to form a tarnish or rust (oxide). Examples of metallic oxides are sodium oxide, Na_2O, aluminum oxide, Al_2O_3, iron(III) oxide (ferric oxide), Fe_2O_3, and copper(II) oxide (cupric oxide), CuO.

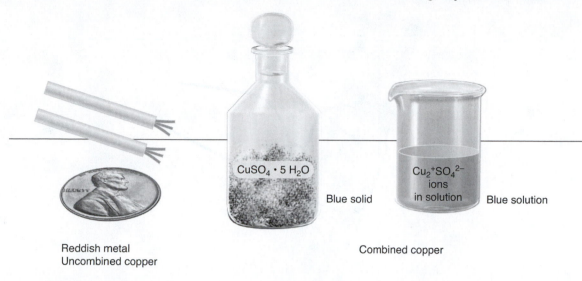

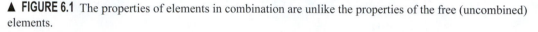

CuSO₄ · 5 H₂O

Blue solid

$Cu_2^+SO_4^{2-}$
ions
in solution

Blue solution

Reddish metal
Uncombined copper

Combined copper

▲ **FIGURE 6.1** The properties of elements in combination are unlike the properties of the free (uncombined) elements.

Replacement Series

In this experiment you will be ranking some of the metals according to their activities, from most to least active. One way to do this is to observe the relative vigor of their reactions (tendency to react) with nonmetals such as oxygen, water, and acid. Another method, which differentiates more clearly, is to note whether one metal can replace another in a chemical compound. The general rule is that the more active metal replaces the less active one. This is easily observed in reactions between metals and metal ions (cations) in solution. Iron, for example, replaces copper from copper(II) chloride, $CuCl_2$, because it is more active than copper (figure 6.2).

$$Fe + CuCl_2 \rightarrow FeCl_2 + Cu$$

In this example two electrons are transferred from the neutral iron atom to the positive copper ion. This is more apparent if the equation is written in net ionic form:

$$Fe + Cu^{2+} \rightarrow Fe^{2+} + Cu$$

This reaction, which involves the transfer of electrons, can be divided into two half-reactions that show exactly where electrons are gained and lost.

$$Fe \rightarrow Fe^{2+} + 2e^- \text{ and } Cu^{2+} + 2e^- \rightarrow Cu$$

The general reaction for the replacement of a metal ion by another metal may be written as follows. (We are arbitrarily showing only one electron per atom being transferred.)

$$M_A + M_B^+X^- \rightarrow M_A^+X^- + M_B$$

| Free metal A | Salt of metal B | Salt of metal A | Free metal B |

Or, in net ionic form,

$$M_A + M_B^+ \rightarrow M_A^+ + M_B$$

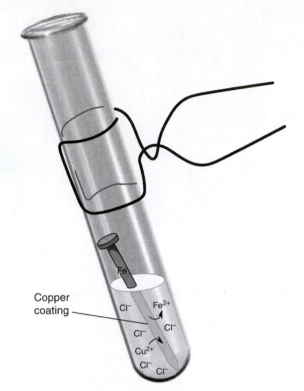

Copper
coating

▲ **FIGURE 6.2** The more active iron replaces the less active copper from solution.

Replacement of Hydrogen from Acids and Water

Acids are an important group of compounds, which, in water, produce hydronium ions,[3] H_3O^+.

Hydronium ions enter into replacement reactions with metals; so hydrogen appropriately is included in an activity or replacement ranking among metals. Those metals more active than hydrogen will replace hydrogen from acids, but those less active are unable to do so.

$$Zn + 2\,HCl \rightarrow ZnCl_2 + H_2\,(g)$$

Or, in ionic style,

$$Zn + 2\,H^+ \rightarrow Zn^{2+} + H_2.$$

You may consider water to be a special case of a very weak acid. Metals that replace hydrogen from acids also liberate hydrogen from water, but drastic conditions may be required. Only the most active metals, such as sodium and potassium, replace hydrogen from water readily at room temperature. The reaction of sodium with water is one you will observe today. This vigorous reaction can be sufficiently exothermic to ignite the explosive hydrogen gas, so you will use only a small amount of sodium. Some reactions of metals with water are

$$2\,Na + 2\,H_2O \xrightarrow{\;25°C\;} 2\,NaOH + H_2,$$

$$3\,Fe + 4\,H_2O \xrightarrow[\text{red heat}]{800°C} Fe_3O_4 + 4\,H_2.$$

---

[3]Protons (H^+ ions) released by acids exist in water solution in hydrated form as hydronium ions, H_3O^+, or even $H(H_2O)_n^+$.

These ions are often represented by H^+ for simplicity and convenience.

Metal Activities and the Periodic Chart

The activity of a metal is related to the ease with which its outer electrons can be removed. Generally, atoms having the fewest electrons in the outer shell and those with the largest radii lose their electrons most easily. In the periodic chart, elements with the fewest outer electrons are located toward the left side, and those with the largest radii are found toward the bottom. Therefore, the most active metals are located toward the left side and lower left corner of the chart.

ENVIRONMENT, CULTURE, AND CHEMISTRY
Recycling Scarce Metals

Except for calcium and aluminum in soils and clays and sodium in the seas, most metals are quite scarce in the earth's crust. It has been possible to mine rich metal ore deposits for centuries but they are getting scarcer and leaner. As established and developing societies use these deposits at a fast rate, metals are becoming less plentiful and more expensive. What can be done? Fortunately metals can be recycled relatively easily by collecting and remelting them. Scrap iron (old cars for example) has been used for years, and more recently, aluminum cans are being collected for recycling, thus saving huge amounts of electrical energy required to process aluminum from its ore. As metals become more precious, we will use less and recycle more. In theory, recycling of metals can approach 100% so there is no real reason to run out of metals–at a price. Why isn't more recycling done right now? The answer is cost. If metals were as expensive as gold, everyone would do it voluntarily. At the current prices for metals, it has been politically unrealistic to require the additional expense of near total recycle. Nearly everyone wants to conserve and protect the environment provided that the price is minimal. To make a major impact the cost would not be minimal and it would affect the standard of living. The debate centers on cost vs. benefit. Developing countries especially tend to see the cost of environmental progress as too great. They want what the richer countries have: new cars and more material goods.

Pre-Laboratory Questions | 6

1. From your previous knowledge of the tendencies of certain metals to rust or tarnish, and from your reading in this experiment, select at least three metals from the following list which are quite active, three which are moderately active, and three which are quite inactive: iron, aluminum, tin, gold, sodium, calcium, silver, platinum, nickel, potassium, copper, lead, and magnesium.

   a. Quite active

   b. Moderately active

   c. Quite inactive

2. Name a use for each metal: chromium, nickel, zinc, radium, tantalum, mercury, tin, and silver.

   a. Chromium   e. Tantalum

   b. Nickel   f. Mercury

   c. Zinc   g. Tin

   d. Radium   h. Silver

3. Describe observable differences between elemental copper and Cu^{2+} ion and between H_2 and $H^+(H_3O^+)$ ion.

4. Describe what happens when a metal reacts with water or acid.

EXPERIMENT

1. Reaction of Metals with Oxygen

In this part of the experiment, you will make a variety of observations on the reactivity of oxygen in the air with several metals. Record your observations in the Table for this Section (1) and for Sections 2 and 3; answer the questions that follow on the Observations and Results sheet.

A. Reaction at Room Temperature

DEMONSTRATION: Your instructor will place small strips of metals on a sheet of paper and label them. Each metal will be scratched to expose a clean surface to the air. Suggested metals are iron, tin, lead, copper, magnesium, aluminum, and zinc. Observe the metals throughout the laboratory period. Do they react with oxygen, that is, do they become tarnished?

Your instructor will cut a piece of sodium. Notice how easily sodium can be cut with a knife. Observe a freshly cut piece and see what happens when air comes into contact with the metal. Does a reaction occur between sodium and oxygen in the air? (Answer Question 1A, 1–3 and fill in the table on the Observations and Results sheet.)

Pour this waste down the drain with plenty of water.

B. Reaction at Elevated Temperature

Most of the metal samples, which you placed on the paper towel, from Part A, which your instructor has placed on the sheet of paper, probably have not reacted noticeably with air. Reactions that are sluggish at room temperature often can be accelerated by raising the temperature. In those cases where a reaction did not occur, except for iron, magnesium, and zinc, obtain small samples of the metals and heat in a clean, dry test tube. Cautiously hold iron, magnesium, and zinc in the flame with tongs. Do any of the metals burn (react) with oxygen? Could you suggest an ingredient for making flares? (Answer Question 1B, 1–8 and fill in the table.)

C. Effect of Surface Area

To determine whether the amount of surface area has an effect on the rate of reaction, use a burner to heat a bulky piece of iron, such as a nail, and also a very small sample of iron with a large surface area, such as steel wool. Which of the two reacts more readily with oxygen? After burning in air, the steel wool may still hold its shape. Take some between your fingers when it has cooled. Is it still metallic iron? (Answer Question 1C, 1–4.)

Allow any unburned residue to cool and place all leftover metals in appropriate waste container.

2. Reaction of Metals with Water

DEMONSTRATION: Your instructor will cautiously add a small piece of sodium to water in a beaker. (Answer Question 2a and fill in the table.)

To determine whether the metals of part 1 react with water, place a piece of each, **the size of a match head**, in 1 mL of water in a test tube. Remember that water acts as an extremely weak acid and that the evidence for a reaction is the evolution of hydrogen gas (bubble formation). You may wish to let these sit for a few minutes to be sure that you don't miss a slow reaction. (Answer Question 2b and fill in the table.)

3. Reaction of Metals with an Acid

Now repeat the reaction of part 2 with three different concentrations of a strong acid, such as hydrochloric acid, HCl. The concentrations are 1 *M*, 6 *M* and 12 *M*; 6 *M* is six times more concentrated than 1 *M* and 12 *M* is twice as concentrated as 6 *M*.

First place a small piece of each metal which did **not** react with water **(but never sodium!)** in 1 mL of dilute (1 *M*) HCl in test tubes. Observe the rates of reaction (evolution of hydrogen gas). Those metals which cause the evolution of hydrogen gas can be presumed to be more active than hydrogen.

Slow reactions often can be speeded by increasing the concentration of a reactant. If any of the metals reacts extremely slowly with the 1 *M* acid, or fails to react, test its reactivity in 6 *M* hydrochloric acid, HCl. Finally, try any metal in 12 *M* HCl if it fails to react in 6 *M* HCl. Record in the table the data on the reactivity of each metal. Indicate if 6 *M* or 12 *M* HCl was required and what you observed. The data from all three combinations of acid must be recorded in the column under the "With Acid" heading. (Answer Question 3, a–c and fill in the table.)

4. Reactions of Metals with Other Metal Ions

Compare the reactivity of iron, copper, and silver in the following ways. In separate test tubes place 5 mL of each of the specified salt solutions containing the metal cation, and add a piece of the indicated metal (a strip or a bright nail). Allow at least 15 minutes for a reaction to occur. The more active metal will replace the less active one from its solution. If a replacement reaction occurs you will easily recognize it by the deposition of the new metal on the nail or strip. The metals and salt solutions are:

>   iron (nail) in 0.1 *M* $CuSO_4$ solution,
>   copper (strip) in 0.1 *M* $FeSO_4$ solution,
>   copper (strip) in 0.1 *M* $AgNO_3$ solution.

(Answer Question 4, a–c.)

> **Upon completing parts 2 through 4, place all unused metals in the appropriate container and pour all solutions down the drain.**

QUESTIONS AND PROBLEMS

1. Using the table on the activities of metals (appendix, table A.2), arrange the following elements from most active to least active: potassium, lead, sodium, barium, calcium, magnesium, aluminum, and tin.

2. Locate each of the metals in the previous question on a periodic chart, and make some general observations: Are the more active metals located toward the right or the left side? Toward the top or the bottom of the chart? Where are the nonmetals in relation to the metals? Based on the chart, which is the most active metal?

3. When metal oxides are dissolved in water they form hydroxides, which are basic. The most active metal oxides make the strongest bases. Write, in order of basic strength, from weakest to strongest, the formulas for the hydroxides of each of the metals studied in this experiment.

4. Metals form tarnishes not only with oxygen but with other nonmetals and their compounds. Silver is a noble metal which remains shiny in pure air but turns black when it comes into contact with fumes from certain mineral springs or with egg yolk. What is the formula for the tarnish? (Hint: Recall that hydrogen sulfide, H_2S, called rotten egg gas, is found in mineral springs.)

5. Since prehistoric man found his first gold nugget, it is estimated that some 40,000 tons of gold (within an error of 10%) has been taken from the earth, most of which is still on hand. If all of this gold were to be collected and shaped into a cube, what would be its dimension in cubic feet? Gold is 19.3 times denser than water, and one cubic foot of water weighs 62.4 lbs. Would this volume of gold fit inside the building you are now occupying?

EXPERIMENT

Getting Acquainted with Metals

6

OBSERVATIONS AND RESULTS

Table for Questions in Sections 1, 2, and 3: Reactions of Metals with Oxygen, Water, and Acid

In the table below record your observations on the reactivities of the metals used in the experiment. Use qualitative terms such as *fast, slow, vigorous, sluggish,* or *no reaction* to describe what happened. Make a separate record for each concentration of acid.

Metal	With Air at Room Temperature	Heated in Air	With Water	With Acid
_____	_____	_____	_____	_____
_____	_____	_____	_____	_____
_____	_____	_____	_____	_____
_____	_____	_____	_____	_____
_____	_____	_____	_____	_____
_____	_____	_____	_____	_____
_____	_____	_____	_____	_____
_____	_____	_____	_____	_____
_____	_____	_____	_____	_____
_____	_____	_____	_____	_____
_____	_____	_____	_____	_____

1A. Reaction at Room Temperature

1. Did you observe any tarnishing on the surface of the freshly cut sodium? If so, what is the name of the corrosion product? Was sodium metal a reactant?

2. Write an equation for the reaction which occurred on the surface of the sodium metal.

3. Why is sodium normally stored under a liquid such as toluene or mineral oil?

1B. Reaction at Heated Temperature

1. Write equations for the reactions that occurred when you heated zinc and aluminum in the presence of oxygen.

2. What is the formula of the compound that was formed when magnesium was heated in air?

3. Make a sketch showing valence electrons for an atom of the free magnesium metal and for the magnesium ion in the resulting compound.

4. Did the atoms of magnesium lose electrons when the sample was heated?

5. Did other atoms accept electrons from magnesium? What kind of atoms?

6. Did metals react with oxygen more rapidly at flame temperature than at room temperature? Can you explain your observation?

7. Would any of the burning metals make a good night flare?

8. Would any of these metals be expected to react differently if the atmosphere were pure oxygen rather than the 20% oxygen in air? Explain.

1C. Effects of Surface Area

1. Which type of iron sample has the greater surface area?

2. Which type of iron reacted more rapidly with oxygen? What is your evidence?

3. After heating, do the observed properties suggest that the steel wool is still metallic iron?

4. Write an equation showing the oxidation (rusting) of iron to iron(III) oxide.

2. Reaction of Metals with Water

a. Describe the reaction of sodium and H_2O.

b. Write equations for sodium and any of the other metals that reacted with water.

3. Reaction of Metals with an Acid

a. Write equations for the reactions (if any) that occurred between hydrochloric acid and aluminum, and hydrochloric acid and copper.

b. Of the metals you have examined in this part of the experiment, which would be the most suitable for use in jewelry? Which would be the least suitable?

c. Zinc is more active than copper. *On the basis of this statement only*, which of the following statements are true?

_____ Zinc has a greater tendency than copper to exist in the combined form (as a cation).

_____ It is easier to get copper metal from copper sulfate than to get zinc metal from zinc sulfate.

_____ Zinc tarnishes faster than copper.

_____ Zinc replaces hydrogen from acids but copper does not.

4. Reactions of Metals with Other Metal Ions

a. What, if anything, did you observe in each of the following trials? Write an equation for the reaction if one occurred.

Iron with $CuSO_4$ solution:

Copper with $FeSO_4$ solution:

Copper with $AgNO_3$ solution:

b. When a metal atom becomes an ion, or an ion becomes an atom, there is a transfer of electrons. In the preceding reactions, draw circles around the metal atoms which lost electrons and boxes around the metal ions which gained them (reactants).

c. Which is the more active: (a) iron or copper, (b) iron or silver, (c) silver or copper? Arrange the three metals in order of decreasing activity.

The Halogen Family: Some Colorful Nonmetals | 7

OBJECTIVES

1. To become acquainted with the physical and chemical properties of the halogens, the most active family of nonmetallic elements, and their compounds.

2. To learn about some important consumer products which contain members of the halogen family.

3. To relate the information from this experiment to such questions as the medical problem of goiters and the political controversy over the fluoridation of drinking water.

DISCUSSION

The odor of chlorine escaping from a treated swimming pool is familiar to nearly everyone. In pure form, chlorine is a yellow-green, toxic gas. In 1965, when a barge, loaded with liquefied chlorine, sank in the lower Mississippi River, many of the surrounding cities were evacuated because of the threat from the escaping gas. In World War I allied troops were terrorized by attacks in which chlorine and other poison gases were used.

Chlorine and the somewhat less familiar fluorine, bromine, and iodine compose the colorful (the elements are colored and the reactions spectacular), interesting, and useful family known as the halogens. Your grandparents were probably more familiar with elemental iodine than you are. Tincture of iodine, an alcoholic iodine solution, and Merthiolate were widely used for the treatment of cuts before the introduction of antibacterials such as Bactine and Neosporin.

In several later experiments you will become acquainted with elemental bromine, a reddish brown, fuming liquid, which is best handled as a dilute solution. Bromine is one of three elements which are liquids at room temperature. Can you name another? Fluorine, like chlorine, is a yellow-green gas. You will not be introduced to it in the laboratory, because fluorine is too dangerous to be studied without elaborate safety precautions (to say nothing of the need for experience). Fluorine is the most reactive of the nonmetals.

Reactions of the Elemental Halogens

Like most groups in the periodic chart, members of the halogen family have similar chemical properties. Each of them has seven electrons in the outer shell. Consequently, they all react to complete an outer octet of electrons by gaining an electron, either by accepting one from a metal (ionic bond formation) or by sharing a pair with a nonmetal (covalent bond formation). This tendency to complete an outer octet is so strong that the elemental form of the halogens is molecular; that is, the halogens exist as diatomic molecules $(F_2, Cl_2, Br_2, \text{and } I_2)$ rather than as isolated atoms. Diatomic molecules incorporating two different halogens are also known, for example, ICl. The tendency of the halogens to react with a metal to gain an extra electron is also strong. This is why the free elements are so reactive, corrosive, and toxic.

The reactions of chlorine with sodium (a metal) and with hydrogen (a nonmetal) can be diagramed using electron dot symbols.[1]

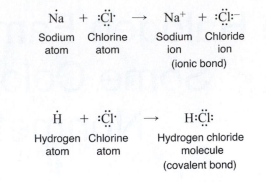

Compounds of the Halogens

The affinity that a halogen has for an electron largely accounts for the stability of the anion. The univalent anion, once formed, does not want to give up its extra electron. Therefore, fluoride, F^-, chloride, Cl^-, bromide, Br^-, and iodide, I^-, ionic compounds are relatively unreactive and comparatively non-toxic. We eat sodium chloride, NaCl, as table salt, but chlorine gas, Cl_2, is extremely toxic and destroys tissue. Fluorine, F_2, is too reactive for you to use, yet stannous fluoride, SnF_2, is sufficiently harmless in small concentrations to put in toothpaste. You will want to keep in mind these important differences in properties between an element and a compound as you do today's experiment. Recall that this same distinction exists in the case of an anion in metals and their cations (experiment 6).

The halogens appear in a number of complex ions, typified by sodium hypochlorite, NaOCl. Here the chlorine is covalently bonded to oxygen, and the negative charge is on the hypochlorite ion, OCl^-. Other complex ions in which halogens are covalently bonded to oxygen are chlorate, ClO_3^-; bromate, BrO_3^-; periodate, IO_4^-; and perchlorate, ClO_4^-. Compounds of halogens with carbon will be studied in later experiments.

Test for the Halides

Certain of the reactions of the halide ions in solution can serve as tests for their presence. You will use these in today's experiment. For example, chloride, bromide, and iodide ions react instantly with silver ion, Ag^+, in nitric acid solution to form a precipitate.[2] This is illustrated with the reaction of sodium chloride, NaCl, and silver nitrate, $AgNO_3$.[3]

$$\underset{\text{(Soluble)}}{NaCl} + \underset{\text{(Soluble)}}{AgNO_3} \rightarrow \underset{\text{(Insoluble)}}{AgCl} + \underset{\text{(Soluble)}}{NaNO_3}$$

---

[1]At sufficiently high temperatures chlorine, hydrogen, and sodium do exist as single atoms.

[2]If the solution were not acidic, silver would precipitate with CN^-, CO_3^{2-}, OH^-, PO_4^{3-}, SO_4^{2-}, and certain other ions as well.

[3]When this equation is written in ionic style, it shows clearly the essentials of the chemical reaction.

$Na^+ + Cl^- + Ag^+ + NO_3^- \rightarrow AgCl + Na^+ + NO_3^-$

Or, leaving out the spectator or nonparticipating ions (net ionic style),

$Cl^- + Ag^+ \rightarrow AgCl$.

Silver chloride, AgCl, is white; silver bromide, AgBr, and silver iodide, AgI, have slightly different shades of yellow and are difficult to distinguish from each other. Silver chloride, AgCl, is identified as the only one of the three that dissolves in 6 M ammonium hydroxide, NH_4OH (ammonia water, NH_3), with the formation of the complex silver ammonia ion.

$$AgCl + 2\,NH_3 \rightarrow Ag(NH_3)_2{}^+Cl^-$$

Each of the halogens is displaced from its salt by a more active halogen (located above it in the periodic chart). The reactions of sodium bromide and sodium iodide with elemental chlorine, Cl_2, will illustrate this.

$$Cl_2 + 2\,NaBr \rightarrow Br_2 + 2\,NaCl$$

$$Cl_2 + 2\,NaI \rightarrow I_2 + 2\,NaCl$$

In these reactions, which are excellent tests for the presence of bromide or iodide ions, a chlorine solution (normally aqueous) will produce the reddish brown color of bromine or the purple color of iodine if the corresponding ion is present. You will use the reaction of chlorine with iodide to confirm the presence of iodide in commercial iodized table salt. As you observe the reaction in your test tube, try to visualize the action between the particles. The more active chlorine atom succeeds in removing an electron from the iodide ion.[4]

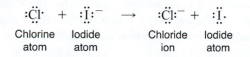

| Chlorine atom | Iodide atom | Chloride ion | Iodide atom |

---

[4]This equation has been simplified. Think of chlorine originally as the diatomic Cl_2. Any iodine formed becomes I_2.

ENVIRONMENT, CULTURE, AND CHEMISTRY
Chlorination of Drinking Water: A Necessity and a Hazard

Chlorination of drinking water in our cities is an essential process because without the presence of a good disinfectant various deadly pathogens such as typhoid, cholera, or dysentery would spread rapidly, causing devastating epidemics. In contrast, homes in rural areas because of their isolation, may obtain quality drinking water directly from wells. Large cities, however, definitely must rely on disinfection of their drinking water because of crowded living conditions, which could allow the rapid spread of disease. The common chlorine disinfectants are hypochlorite ion, OCl^-, hypochlorous acid, $HOCl$, the chloramines, NH_2Cl and $NHCl_2$, and small amounts of molecular Cl_2. In addition to destroying pathogens at the treatment center, it is imperative that the disinfectant remain in the drinking water until it reaches the final location since pathogens are easily transferred from the surrounding soil as the water travels through the city water mains. In fact water mains and sewer lines are frequently located close together and, therefore, transfer between the two often occurs. Chlorination of drinking water is not without a serious downside, however, since chlorine disinfectants react with humic materials (from decomposed trees and plants) and other organic compounds present in the pre-treated drinking water, producing toxic and carcinogenic contaminants (by-products).

Chloroform, a carcinogen, is a common contaminant, but there are other serious chemicals, many of which have not even been identified. Filters at the treatment plant and in the home can remove some, but not all, of the various contaminants. Also, in the shower or bathtub the contaminants are readily absorbed through the skin or inhaled from the warm water. Although they are a definite hazard, modern society appears to be saddled with chlorine disinfectants since a suitable substitute has not yet been developed.

Pre-Laboratory Questions | 7

1. Indicate the charge on each ion in each of the following halogen compounds: $AgCl$, BiI_3, and $CsBr$.

2. Showing only valence electrons, write an equation for the formation of an iodine molecule from two iodine atoms.

3. Showing only valence electrons, write an equation for the reaction that occurs between sodium, Na, and fluorine, F_2. What kind of bond is formed in this reaction?

4. Write balanced equations for (a) the test reaction for chloride with silver ion and (b) the reaction of sodium iodide with elemental bromine.

5. Write the formulas for the following ionic compounds: potassium chloride, magnesium chlorate, calcium hypochlorite, and iron(III) perchlorate.

6. What disease results from the absence of iodide in the diet? What are the symptoms of this disease?

EXPERIMENT

1. Physical Properties of the Halogens and their Anions the Halides

A. Comparisons of the Solubilities of Iodine and Sodium Iodide

With a spatula, place a few crystals of iodine, I_2, in each of two dry test tubes. To one test tube add 1 mL of water and shake. Is I_2 soluble? Partially soluble? To the other test tube, add 1 mL of methylene chloride, CH_2Cl_2, and shake. Repeat the solubility determinations on sodium iodide, NaI. (Record your observations in 1A on the Observations and Results sheet.)

> **Dispose of all aqueous and organic solutions (methylene chloride) in separate appropriate containers.**

B. Effect of Heat on Iodine and Sodium Iodide

Heat a small amount (the size of a matchhead) of sodium iodide, NaI, on a clean spatula for 2 to 3 minutes, and note whether it melts or evaporates or does both. Repeat the procedure with I_2. (Record your observations and answer Question 1B.)

2. Synthesis of a Halogen: Chlorine

In this part of the experiment, you will synthesize molecular chlorine, Cl_2, by the same reaction that occurs in chlorinated swimming pools. To 1 mL of commercial swimming pool hypochlorite solution[5] in a test tube add (very cautiously) 2 drops of 36% (4.6 M) sulfuric acid, H_2SO_4. Cautiously waft the vapors from the test tube to determine if Cl_2 is present. If you are unable to detect the odor of chlorine, add more drops of sulfuric acid until you can. (Small amounts of chlorine will not hurt you.) Can you see gaseous, pale yellow Cl_2 in the test tube above the liquid? Now wet a piece of starch-iodide paper with water and place it in the mouth of the test tube so it will be exposed to the vapors, and note what happens. A blue-black color is further confirmation of the presence of Cl_2.

After you detect the presence of chlorine, add 1 mL of CH_2Cl_2 to the test tube and continue to add sulfuric acid by drops, with constant stirring, until the color of Cl_2 is apparent in the methylene chloride layer. Cl_2 is much more soluble in methylene chloride than in water. (Answer Question 2, a–i.)

> **Dispose of waste in the appropriate container and the starch paper in the garbage can.**

3. Precipitation with Silver Nitrate: Test for the Halides

Place approximately 0.2 g of sodium chloride, NaCl, in a test tube and dissolve it by adding 2 mL of distilled water. Similarly prepare test tubes containing solutions of sodium bromide, NaBr, and sodium

---

[5]Bleach solution is an aqueous mixture of sodium hypochlorite, NaOCl, and sodium chloride, NaCl, made according to the reaction, $Cl_2 + 2\,NaOH \rightarrow NaOCl + NaCl + H_2O$.

iodide, NaI. Now add drops of dilute 1% silver nitrate, $AgNO_3$, solution to each, and note the color of the silver halide that precipitates. Verify the presence of silver chloride, AgCl, by stirring in drops of 6 *M* ammonium hydroxide, NH_4OH, until the precipitate dissolves. (Answer Question 3, a–d.)

Use the AgCl test to detect chloride ion, Cl^-, in the saliva and tap water. To a few drops of saliva and an equal amount of distilled water in a clean test tube, add a few drops of the dilute $AgNO_3$ solution. Does a precipitate form? Record your observations. Now repeat the AgCl test on tap water to determine if chloride ion is present. (Answer Question 3, e–f.)

Commercial iodized table salt, which many people use every day in their food, is primarily sodium chloride with a trace—approximately 10 parts per million (ppm)—of potassium iodide, KI. The presence of this additive in salt provides a convenient way to ingest the iodide necessary to prevent goiter, a disease of the thyroid gland. On the basis of the colors of the silver halide precipitates which you observed earlier, see if you can detect the small amount of KI in the table salt. Conduct the test, as previously, by dissolving a small amount of table salt in distilled water and adding drops of $AgNO_3$ solution. Based on the color of the precipitate, are you able to detect the KI? (Answer Question 3g.)

4. Reaction of a Halogen with a Halide Ion

A more sensitive method for detecting potassium iodide in iodized salt, involves using chlorine to convert the iodide to molecular iodine. (Refer to the discussion section for the equation.) Since the amount of I_2 that is formed is extremely small, it will be difficult to observe the faint purple color of I_2 directly. However, starch forms an intense blue-black complex with very small amounts of I_2. You should be able to detect this blue-black (purple) complex. To become familiar with this test, dissolve a few crystals of reagent KI in 2 mL of distilled water in a test tube and add 5 drops of starch indicator solution. Mix thoroughly. Then add chlorine water (Cl_2 dissolved in water), drop by drop, shaking after each drop, until a color develops. Note the color of the solution.

Now repeat the test, using iodized table salt and a larger test tube to dissolve more salt. In a 15-cm test tube mix 5.0 g of the salt with 5 mL of water and shake vigorously to dissolve as much of the salt as possible. Add 5 drops of starch indicator solution and again stir thoroughly Then add chlorine water 1 drop at a time, shaking *immediately* and vigorously after each drop, and note the color that develops. The purple color may be faint and fleeting. Is KI present in iodized table salt? You should expect the color of the solution to be fainter with the table salt than with pure KI. (Answer Question 4, a–c.)

Dispose of all waste from parts 3 and 4 of this experiment in the appropriate containers.

QUESTIONS AND PROBLEMS

1.   The molecular structure of the important hormone thyroxine is

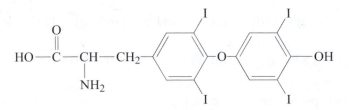

Thyroxine is synthesized (produced) by the thyroid gland. On the basis of the kinds of atoms that are present in a thyroxine molecule, explain why iodide ion (as KI) is required in the diet.

2.   Fluoridation of water involves addition of 1 ppm of sodium fluoride, NaF, to public drinking waters for the purpose of reducing tooth decay. The opponents of the fluoridation of water sometimes mention "the dangers of the yellowish green gas fluorine." What have you learned in this experiment that would provide you with evidence against this particular line of argument? (Hint: Recall the difference, in both electronic structure and properties, between the halogen and the halide ion.)

3.   The common pain reliever Bromo Seltzer once contained bromides. They have now been replaced by other chemicals because the bromides caused harmful side effects. Describe how you would demonstrate that bromide is no longer present in Bromo Seltzer.

	EXPERIMENT
The Halogen Family: Some Colorful Nonmetals	**7**

OBSERVATIONS AND RESULTS

1. Physical Properties of the Halogens and Their Anions, the Halides

    A. *Comparisons of the Solubilities of Iodine and Sodium Iodide*

        Indicate whether soluble, partially soluble, or insoluble.   *I_2*   *NaI*

        In water   _____   _____

        In methylene chloride   _____   _____

    B. *Effect of Heat on Iodine and Sodium Iodide*

        Did it melt?   _____   _____

        Did it vaporize?   _____   _____

        What does the difference in the behavior of these compounds when they are heated suggest about the attraction between the particles (ions or molecules) of each compound in the solid state?

2. Synthesis of a Halogen: Chlorine

    a. What is the trade name of the commercial swimming pool hypochlorite solution?

    b. What is the percentage of sodium hypochlorite in this solution?

c. Describe the odor of chlorine.

d. Were you able to observe the color of gaseous chlorine? If not, why not?

e. Did the starch-iodide paper turn dark?

f. What was the color of Cl_2 in methylene chloride?

g. Why was the color of gaseous chlorine less intense than the chlorine-methylene chloride solution?

h. Give two important functions of chlorine in a swimming pool.

i. Complete and balance the following equation which describes the synthesis of chlorine.

$$NaOCl + NaCl + H_2SO_4 \rightarrow \qquad + \qquad + H_2O$$

3. **Precipitation with Silver Nitrate: Test for the Halides**

   a. Describe the color of each of the silver halides AgCl, AgBr, and AgI.

   b. On the basis of this experiment, can $AgNO_3$ be used to determine whether an unknown salt is a chloride, a bromide, or an iodide? Explain.

   c. Write an equation for the formation of each of the silver halides from their respective sodium salts and $AgNO_3$.

   d. Is silver chloride, AgCl, soluble in 6 M NH_4OH?

   e. Did precipitates form when $AgNO_3$ was added to saliva? Tap water?

   f. Do these observations confirm the presence of Cl^- in saliva and tap water? Explain.

   g. Judging by the color of the precipitate that formed when you added $AgNO_3$ to table salt, were you able to detect KI in table salt? What is the evidence for your answer?

4. Reaction of a Halogen with a Halide Ion

a. What was the color of the starch-iodine complex formed from reagent KI and Cl_2? Write an equation for the reaction of KI and Cl_2.

b. Were you successful in detecting KI in iodized table salt, using chlorine water and the starch indicator solution? On what observations do you base your conclusion?

c. Again, state why KI is added to table salt.

Analysis of Commercial Bleaches: A Comparison of Two Competing Products | 8

OBJECTIVES

1. To gain an understanding of chemical procedures which can be used to analyze quantitatively some important consumer products.

2. To become familiar with the technique of titration.

3. To assess critically the advertising claims for bleaches and to determine if such claims are supported by scientific data.

DISCUSSION

Anyone familiar with television advertising is certain to be aware of the competition in recent years among the manufacturers of household bleaches. Each advertiser seems to suggest that its product is stronger and more effective than the competitors' products. In this experiment you will analyze two liquid bleaches (Clorox ©, or another name brand and store brand) to determine for yourself whether there is a sound basis for the advertising claims. When you read the labels you will see that the active ingredient, sodium hypochlorite, $NaOCl$, is the same in both products. The ion responsible for the bleaching action is the hypochlorite ion, OCl^-. If there is a real difference between the two commercial bleaches, it must be ascribed to differences in concentration of the active ingredient. Your task will be to determine the concentration (strength) of sodium hypochlorite in each bleach and thus to evaluate the advertising claims.

Hypochlorite Chemistry

In order to understand the analysis procedure, we must probe somewhat deeper into the chemical reactions of halogen compounds. *Sodium hypochlorite,* $NaOCl$, reacts with iodide ions in sodium iodide, NaI, in the presence of acetic acid to form iodine.

$$NaOCl + 2KI + 2HC_2H_3O_2 \rightarrow I_2 + NaCl + 2KC_2H_3O_2 + H_2O$$

Iodine, in turn, reacts with aqueous sodium thiosulfate, $Na_2S_2O_3$.

$$I_2 + 2Na_2S_2O_3 \rightarrow Na_2S_4O_6 + 2NaI$$

These reactions provide a simple way of finding the amount of hypochlorite present in bleach. As a solution of sodium thiosulfate is added from a buret to the solution containing iodine, the end point (completion of the reaction) is signaled by the disappearance of the iodine color.[1]

Adding a solution of one reactant of known concentration (a standard solution) to a solution of another reactant of unknown concentration until the end point is reached is called *titration*. You will use the titration procedure in today's experiment.

The percentage of sodium hypochlorite in the bleach is found by multiplying the volume of sodium thiosulfate by a constant, 0.340.

$$\text{Percentage of NaOCl} = 0.340 \ V$$

where V is the volume in milliliters of sodium thiosulfate solution used in the titration. The constant in the equation has been calculated assuming a density of 1.094 g/mL for the bleach solution and a concentration of 0.10 mole/L for the sodium thiosulfate solution.

Titration Techniques

Read carefully the pertinent portions of experiment 14 on the use of a buret and a pipet. A summary of the correct procedure is given here.

1. Never draw liquids into the pipet with your mouth—use a rubber bulb.

2. Before use, clean the glassware with soap and water, and then rinse it thoroughly with distilled water.

3. Rinse the buret and pipet with a small amount of the liquid which will be used in the titration so that the liquid film remaining on the glass walls after washing will not dilute your solutions.

4. Allow enough of the liquid to run through the tip of the buret to dispel all air bubbles.

5. Read the bottom of the meniscus.

6. Record the volume to two decimal places.

---

[1]Usually a starch indicator, which forms a blue-black complex with iodine, is added just before the completion of the reaction to accentuate the end point. Your instructor will tell you if you are to use a starch indicator.

Pre-Laboratory Questions | 8

1. What is an indicator? What indicator is used in this experiment?

2. What is meant by the end point of a reaction?

3. What is a standard solution? Which is the standard solution for today's experiment?

4. Write the equation for the titration of iodine, I_2, with sodium thiosulfate, $Na_2S_2O_3$.

5. If the most reactive halogen (one which can replace the ion of another—see experiment 7) is the most effective, which of the following would make the best bleach: Cl_2, Br_2, or I_2?

EXPERIMENT

Prepare to determine the concentrations of sodium hypochlorite in bleach by cleaning and mounting the buret as shown in experiment 14. Dissolve 0.35 g of potassium iodide, KI, in about 50 mL of distilled water in a 250-mL Erlenmeyer flask and add about 1 mL of concentrated acetic acid, $HC_2H_3O_2$. Carefully pipet 1 mL of one of the bleaches into this solution and stir. (The amount of iodine, I_2, formed is equivalent to the sodium hypochlorite originally present in the bleach.) Fill the buret to within a few milliliters of the top with sodium thiosulfate, $Na_2S_2O_3$, read and record the exact volume, and proceed to titrate the iodine solution. Swirl the Erlenmeyer flask while the solution from the buret is added. When the brown color of the iodine becomes lighter, slow down the addition from the buret. Add drops from the buret one at a time and stop when the solution becomes colorless (figure 8.1). Record the final volume of sodium thiosulfate in the buret. The difference in your two readings is the volume used. (Record the data for #1 in the chart provided and carry out the necessary calculations.)

Titrate a second and a third sample of the same bleach. Repeat the procedure on the three samples of the second bleach. For each bleach, average the three volumes of sodium thiosulfate, unless one volume is sufficiently different to be discarded. Determine the percentage of sodium hypochlorite by using the

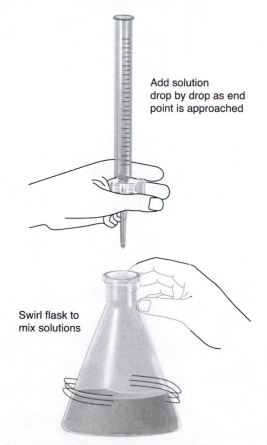

Add solution
drop by drop as end
point is approached

Swirl flask to
mix solutions

▲ **FIGURE 8.1** Approaching the end point of a titration.

equation which was given in the discussion of hypochlorite chemistry. (Record the data for #2 in the Observations and Results sheet provided and carry out the necessary calculations. Answer Questions 3–12.)

Dispose of the titration solutions by pouring them down the drain with ample water.

QUESTIONS AND PROBLEMS

1. Do you have any recommendations which should be incorporated into any new truth-in-advertising legislation? Cite any recent truth-in-advertising legislation of which you are aware.

2. Could you make a judgment as to the effectiveness of commercial liquid bleaches by noting the ingredients on the labels? Explain.

3. At the present time any new prescription drug requires an extensive series of evaluation trials, first with animals and then with patients, to establish efficacy and safety before it is allowed on the market. At the same time hundreds of "natural health supplements" are advertised and sold with no requirement for testing. Is this difference reasonable?

4. Although new varieties of food crops—e.g., tomatoes—are made available to gardeners and farmers each year without food-safety testing, any new transgenic crop must first be put through an expensive testing procedure. Discuss your views on whether this distinction is reasonable or not.

EXPERIMENT

Analysis of Commercial Bleaches: A Comparison of Two Competing Products

8

OBSERVATIONS AND RESULTS

1. Bleach 1 (name) _____

	Run 1	Run 2	Run 3
Buret reading *after* titration	_____ mL	_____ mL	_____ mL
Buret reading *before* titration	_____ mL	_____ mL	_____ mL
Volume of $Na_2S_2O_3$ solution added	_____ mL	_____ mL	_____ mL
Average volume $Na_2S_2O_3$		_____ mL	

Show calculations for determining the percentage of NaOCl in bleach 1.

2. Bleach 2 (name) _____

	Run 1	Run 2	Run 3
Buret reading *after* titration	_____ mL	_____ mL	_____ mL
Buret reading *before* titration	_____ mL	_____ mL	_____ mL
Volume of $Na_2S_2O_3$ solution added	_____ mL	_____ mL	_____ mL
Average volume $Na_2S_2O_3$ solution		_____ mL	

Show calculations for determining the percentage of NaOCl in bleach 2.

3. What is the difference in the strengths of the two bleaches? Is this difference significant?

4. Would either advertiser be justified in claiming that its product is superior to the other?

5. Compare the percentages of sodium hypochlorite which you determined with those on the bleach bottle labels.

	Bleach 1	*Bleach 2*
6. Your percentage	_____ %	_____ %
7. Manufacturer's listed percentage	_____ %	_____ %

8. Ask six other students for the percentages of sodium hypochlorite they calculated for each bleach. Average these percentages and compare the average to the percentages on the bottles.

	Bleach 1	*Bleach 2*
9. Average of six percentages	_____ %	_____ %
10. Manufacturer's listed percentage	_____ %	_____ %

11. Are the averages of the six students' percentages closer to the manufacturers' listed percentages than your own is?

12. Carefully analyze this experiment and point out the areas where you may have made errors.

How Fast Are Chemical Reactions? | 9

OBJECTIVES

1. To observe the wide differences in the rates of various chemical reactions and to discover some factors which affect the rate of a particular reaction.
2. To become familiar with the role of a catalyst in a chemical reaction.
3. To develop competence in using units of concentration.

DISCUSSION

Chemical reactions vary greatly in speed (rate). Some reactions, such as the explosion of methane in a coal mine, are extremely rapid. Others, like geological processes, may be so slow that centuries pass before the reaction has proceeded noticeably. Between these extremes are many reactions of moderate speed that are convenient to study in the laboratory. We try to understand the factors which control reaction rates in order to know how to speed them up or slow them down. We may wish to make certain reactions, such as the cooking of food or the decomposition of wastes, go faster. Others, like the corrosion of iron, certain decay processes, or biological aging, we might like to slow down. The study of reaction rates is called *kinetics*.

Factors Influencing Reaction Rates

Factors which influence reaction rates are found experimentally for each reaction. Some of these variables are concentration of reactants, temperature, catalyst, and surface area (in the case of a reaction taking place at an interface). The concentrations[1] of the various reactants affect the rates of most reactions. Higher concentrations increase the reaction rate, and lower concentrations decrease it. For example, cellular reactions in living organisms, which occur at reasonable rates in air (20% oxygen), become so rapid in a concentrated atmosphere of 100% oxygen that the organism may be destroyed.

Sometimes the concentration affects the rate in a direct linear (one-to-one) relationship: doubling the concentration doubles the rate. Often it is a direct square relationship, in which case doubling the concentration quadruples the rate, and tripling the concentration increases the rate by three squared or nine times. In some instances, a change in concentration has no effect. Each reactant has its individual effect independent of the other reactants. Almost all chemical reactions proceed more rapidly as the temperature is increased. The rate of increase depends on the reaction. For many reactions, especially in organic chemistry, a rule of thumb is that the rate is approximately doubled for each 10°C rise in temperature.

---

[1]Concentrations are usually expressed as molarity, M, which signifies moles per liter. Molarity is the number of moles of dissolved material per liter of solution. This value is obtained by dividing the number of moles of solute by the number of liters of solution. For example, 0.30 mole of H_2SO_4 dissolved in 0.50 L (total volume) would be 0.30 mole/0.50 L or 0.60 mole/L (0.60 M).

A *catalyst* is defined as a substance which affects the rate of a reaction but emerges unchanged from the reaction. Catalysts are usually thought of in a positive sense—as increasing the reaction's speed. Enzymes in the body are examples of some of nature's catalysts. Their presence enables various physiological reactions, such as the digestion of food, to take place at reasonable speeds at body temperature. There are also negative catalysts, called *inhibitors*. An example is an organic amine sold as a rust inhibitor to deter rust formation in automobile radiators.

A good example of a catalyst's influence upon reaction rate is the reaction of permanganate ion, MnO_4^-, with oxalate ion, $C_2O_4^{2-}$, which you will study for today's experiment.[2]

$$2\ MnO_4^- + 5\ C_2O_4^{2-} + 16\ H^+ \xrightarrow{\ Mn^{2+}\ } 2\ Mn^{2+} + 10\ CO_2 + 8\ H_2O$$

The catalyst for a reaction is sometimes indicated above the arrow. Note that Mn^{2+} is both a catalyst and a product in this reaction. Would you expect this reaction to pick up speed as more catalyst is produced by the reactants?

Some reactions, such as the rusting of iron, take place at a surface. The rate of such reactions is dependent on, among other things, the amount of surface area in contact with the other reactants.

Colliding Molecules

You can better understand the factors affecting the rates of reactions by considering the behavior of the individual molecules or particles involved. Most chemical reactions occur because two or more particles (molecules, atoms, ions) collide effectively and bond together or because unstable molecules come apart by breaking bonds. In order for collisions to result in the formation of a new compound, the particles must collide with sufficient force and they must be in proper orientation toward each other. When substances are mixed, particularly in the case of gases or liquids, the number of collisions is astronomically high. Only a fraction of the collisions are likely to be effective. For slow reactions, the fraction of effective collisions is generally much less than for fast reactions.

Changes in conditions that make particle collisions *more frequent* or *more effective* speed the reaction. Thus, the increase in rate owing to higher concentrations of reactants is explained by the increase in the number of collisions. The more particles there are in a given volume, the more they will bump into one another. Raising the temperature causes particles to move faster. An increase in the speed of the particles increases both the *number* and the *effectiveness* of collisions, and so the reaction goes faster. A catalyst helps by providing a more effective pathway or mechanism for the reaction.

The Iodine-Clock Reaction

In this experiment you will study the fascinating reaction between iodate ion, IO_3^-, and sulfite ion, SO_3^{2-}, to form iodide, I^-, and sulfate, SO_4^{2-}.

$$IO_3^- + 3\ SO_3^{2-} \rightarrow I^- + 3\ SO_4^{2-}$$

When the sulfite is used up the reaction is over, and the excess iodate immediately signals the end by reacting with the iodide present to form iodine, I_2.

$$IO_3^- + 5\ I^- + 6\ H^+ \rightarrow 3\ I_2 + 3\ H_2O$$

---

[2]This is the same reaction as that represented by the equation in prelaboratory question 3b. That equation was written in the *molecular complete* form. This one illustrates the convenient *net ionic* form of equation writing, where only the ions and molecules that actually participate in the reaction (not the spectator ions) are shown. The style of equation that most clearly describes a reaction is normally selected.

The appearance of iodine gives this classic reaction a color change so sharp and dependable that it has become known as the "iodine-clock" reaction.[3]

Iodine under these conditions is normally a faint brownish color at first appearance, so an indicator is needed to intensify the color. Therefore, a small amount of soluble starch is added, which combines with the iodine to give an intense blue-black color. The appearance of this color signals the presence of even small quantities of iodine. (See experiment 7.)

You will discover experimentally which factors influence the rate of this reaction. The approach will be to vary some of the conditions, such as the concentrations of reactants and the temperature, and to observe the effects of these changes. When you discover the rules (rate law) that govern the reaction, you can even predict the time required for the reaction under some yet untried set of conditions.

You must be alert and well organized to carry out the experiment successfully. You will be measuring, to the nearest second, the time for the reaction. Each measurement is fairly short. There is time to repeat if you make mistakes. Use this opportunity to develop self-reliance in obtaining meaningful results.

 ENVIRONMENT, CULTURE, AND CHEMISTRY

Oxidation States of the Halogens: Different Uses and Toxicities

This experiment deals with the element iodine in three different states, called oxidation states: IO_3^-, I_2, and I^-. This is a good opportunity to point out that the benefit or danger of a particular substance may depend on its oxidation state as well as specifications such as concentrations and quantities. One cannot just say that a chemical is either good or bad, safe or dangerous; it is often both. The iodate ion, IO_3^-, is a strong oxidizer and can be corrosive and destructive to human tissue. Iodide ion, I^-, in small quantities, is added to ordinary table salt (called iodized salt) to prevent a goiter, which is an enlargement of the thyroid gland. Some areas of our country have sufficient I^- in the soil, which is consequently absorbed in the vegetables that we eat, so that the addition of extra I^- to the diet would not be necessary. The vapor of the diatomic element iodine, I_2, should not be inhaled because of its toxicity and irritation to lung issue. Yet, in small quantities, when dissolved in alcohol, I_2 has been used extensively as a disinfectant for cuts and wounds. Another example of a good and bad chemical is chlorine, Cl. As the free element, Cl_2, a gas, is so toxic that it was used by the German army during World War I as a poison gas. On the other hand, its ion, Cl^-, is not only benign but necessary for the electrolytic balance in the body. But even here, there is the possibility of too much of a good thing since sodium chloride (NaCl) is connected to high blood pressure and water retention. Fluorine, another member of the halogen family, has good and bad qualities, depending on the oxidation state. As the diatomic element, F_2, fluorine is so explosive and toxic that even professional chemists must have special training to use it when pure (undiluted with inert nitrogen gas, for example). In contrast, the fluoride anion, F^-, is used extensively in toothpaste and water supplies (highly diluted) to prevent tooth cavities.

---

[3] These reactions, written in molecular form, including spectator ions, are

$$KIO_3 + Na_2SO_3 \rightarrow KI + 3\,Na_2SO_4$$

$$KIO_3 + 5\,KI + 3\,H_2SO_4 \rightarrow 3\,I_2 + 3\,K_2SO_4 + 3\,H_2O$$

Name ____

Date ____________________ Lab Section ______________________________

Pre-Laboratory Questions | 9

1. Classify the following reactions as fast or slow: (a) the souring of milk, (b) the explosion of dynamite, (c) the digestion of food, (d) the reaction of sodium metal with water, and (e) the precipitation of AgCl when Ag^+ and Cl^- ions are mixed.

   (a)

   (b)

   (c)

   (d)

   (e)

2. If 2 moles of a reagent are present in 8 L of solution, what is the concentration in moles per liter?

3. What observations could you make that would indicate when the following reactions are essentially complete?

   a. the dissolving of marble (calcium carbonate) in hydrochloric acid,

$$CaCO_3 \;+\; 2\,HCl \;\rightarrow\; CaCl_2 \;+\; CO_2 \;+\; H_2O$$

White   Colorless   Colorless   Gas
solid   solution   solution

   b. the reaction of potassium oxalate with potassium permanganate in acid solution,

$$5\,K_2C_2O_4 \;+\; 2\,KMnO_4 \;\; 8\,H_2SO_4 \rightarrow$$

Colorless   Purple   Colorless
solution   solution   solution

$$2\,MnSO_4 \;+\; 6\,K_2SO_4 \;+\; 10\,CO_2 \;+\; 8\,H_2O$$

Colorless   Colorless   Gas
solution   solution

4. What is the function of a catalyst? Name the catalyst which will be used in this experiment.

5. What is the visual evidence which signals the completion of the "iodine-clock" reaction?

EXPERIMENT

1. Fast and Slow Reactions

A. Precipitation of Calcium Carbonate

To 2 mL of 0.1 M calcium chloride, $CaCl_2$, solution in a test tube, add, with stirring, 1 mL of 0.1 M sodium carbonate, Na_2CO_3. Note the result, particularly the speed of the reaction. (Answer Question 1A, 1–3 on the Observations and Results sheet.)

B. Reaction of Permanganate Ion with Oxalate Ion

To 1 mL of 0.1 M sodium oxalate, $Na_2C_2O_4$, solution in a test tube, add, with stirring, 7 drops of 1 M sulfuric acid, H_2SO_4, and then 1 drop of 0.1 M potassium permanganate, $KMnO_4$. Note the time required for the solution to turn from purple to brown and then to become colorless. The reaction may take several minutes, so do not wait. Let the test tube stand, and proceed with part 2. Then come back to this part.

Now add 3 more drops of the $KMnO_4$ solution to the mixture and note the time it takes for the color to disappear. Add 3 more drops of the $KMnO_4$ solution and again note the reaction time as indicated by color disappearance. Mix together a new batch similar to the first, consisting of 1 mL of 0.1 M $Na_2C_2O_4$ and 7 drops of 1 M H_2SO_4, and add 3 drops of 0.1 M manganese sulfate, $MnSO_4$. Now add 1 drop of 0.1 M $KMnO_4$, as before, and note the time required for the color to disappear. (Answer Question 1B, 1–3.)

2. Factors Affecting the Speed of a Reaction: The Iodine-Clock Reaction

You will be making a number of different mixtures containing iodate ion, IO_3^-, and sulfite ion, SO_3^{2-}. The first mixture will be referred to as the *standard mixture*. All others will be variations of this one. You will vary both the concentrations of the reactants and the temperature. Note the reaction time in each case. Then you will be able to evaluate the effect of each of these variables on the rate of the reaction. Bring to your desk about 100 mL of each of the following reagents in clean flasks or beakers and label them solutions A and B.

Solution A. This solution is 0.02 M potassium iodate (4.3 g of KIO_3 per liter of solution).

Solution B. This solution is 0.01 M acidified sodium sulfite-starch (1.3 g of Na_2SO_3, 10 mL of 6 M H_2SO_4, and 5 g of boiled, soluble starch per liter of solution).

Carry out all reactions in a clean 400-mL beaker placed on a piece of white paper. Have a stirring rod available as well as a timepiece that measures seconds.

Mixture 1. This is the *standard run*, called the control, to which the others will be compared. With a clean 10-mL pipet (rinsed out with the solution to be pipetted) measure out 10 mL of solution B into the reaction vessel. Add 80 mL of distilled water, using a graduated cylinder. Stir the solution. Measure out 10 mL of solution A with another pipet and place it in a small container or test tube, so it can be added quickly to the reaction vessel. Pour solution A into the vessel as quickly as possible, with mixing. Stir thoroughly, and measure accurately (to the nearest second) the time interval from the moment of mixing to the color change. Do not be concerned about any droplets of solution A which remain in the container, as solution A is in excess. Repeat the standard run and average the two results.

For the following runs proceed as in the standard run but use the amounts of solutions indicated. Make duplicate runs for mixtures 2, 3, 4, and 5.

Mixture 2. Add 10 mL of solution A to 20 mL of solution B and 70 mL of water.

Mixture 3. Add 10 mL of solution A to 30 mL of solution B and 60 mL of water.

Mixture 4. Add 20 mL of solution A to 10 mL of solution B and 70 mL of water.

Mixture 5. Add 30 mL of solution A to 10 mL of solution B and 60 mL of water.

For mixtures 6, 7, and 8, do not make duplicate runs.

Mixture 6. Add 10 mL of solution A to 10 mL of solution B and 180 mL of water. Expect the iodine-starch color to be less intense than in the standard run.

Mixture 7. In this run the amounts are the same as in the standard run. Before mixing, cool the container holding solution A and the reaction vessel containing water and solution B to 15°C in an ice-water bath. Immediately after mixing, note both the time required for the reaction and the temperature of the mixture.

Mixture 8. Again, the amounts are the same as in the standard run. This time warm both the container of solution A and the reaction vessel containing solution B and water to 45°C before mixing. The color at the end point will be brownish rather than blue at this temperature. (Record data in the table and answer Question 2, 1–8.)

Dispose of all waste from parts 1 and 2 of the experiment in the appropriate containers.

QUESTIONS AND PROBLEMS

1. To retard spoilage, perishable foods are stored in a refrigerator rather than at room temperature. Use your knowledge of reaction rates to explain why.

2. Suggest an explanation for the fact that foods cook faster at higher temperatures.

3. Why do meats cook faster in oil than in water?

4. Name some chemical processes for which inhibitors (negative catalysts) would be desirable. (For instance, how about human aging?)

EXPERIMENT

How Fast Are
Chemical Reactions?

9

OBSERVATIONS AND RESULTS

1. Fast and Slow Reactions

A. *Precipitation of Calcium Carbonate*

1. Write a molecular equation and also a net ionic equation for the reaction of $CaCl_2$ with Na_2CO_3 to form calcium carbonate.

   Molecular equation:

   Net ionic equation:

2. Describe your observations of the reaction.

3. How would you classify the speed of the reaction?

B. *Reaction of Permanganate Ion with Oxalate Ion*

1. How long did it take for the color to change from purple to colorless when the first addition of MnO_4^- was made? How long did it take when the second and third additions were made? Explain any time differences for the three additions.

2. How much time was required for the color to disappear when $MnSO_4$ was present initially? What role is played by Mn^{2+} in this reaction?

3. Could the effect on the reaction have been produced by SO_4^{2-} ion? (Recall that H_2SO_4 was present in all cases in large concentration.) Explain.

2. Factors Affecting the Speed of a Reaction: The Iodine-Clock Reaction

Mixture	Temperature (°C)	Solution A, KIO_3 (mL)	Solution B, Na_2SO_3 (mL)	Water (mL)	Reaction Time (seconds)		
					Run 1	Run 2	Average
1							
2							
3							
4							
5							
6							
7							
8							

1. Compare the relative concentrations of iodate, IO_3^-, and sulfite, SO_3^{2-}, in mixtures 1, 2, and 3.

2. On the basis of your observations of mixtures 1, 2, and 3, make a statement which expresses the relationship between the concentration of sulfite, SO_3^{2-}, and the rate of the reaction.

3. Which mixtures would you compare to determine the effect of the concentrations of iodate, IO_3^-, on the reaction rate?

4. On the basis of your answer to the previous question, examine those results and make a statement that expresses the relationship between the concentration of IO_3^- and the rate of the reaction.

5. Compare the concentration of iodate, IO_3^-, in solution 1 to that in solution 6. Make a similar comparison of the concentrations of the sulfite, SO_3^{2-}, in solutions 1 and 6.

6. How does the rate of the reaction in mixture 6 compare with the standard? Is this observation in agreement with what you would have predicted based on answers to the second and fourth questions? Explain.

7. How did the changes in temperature in mixtures 7 and 8 affect the rate of the reaction?

8. Predict how long it would take for a color change to occur if you started with 20 mL of A, 20 mL of B, and 60 mL of water at room temperature.

Chemical Reactions Can Go Forward and Backward | 10

OBJECTIVES

1. To become aware that many chemical reactions are reversible.
2. To learn about equilibrium and some of the factors which cause equilibria to shift.

DISCUSSION

Occasionally an observant beginning chemistry student will notice that some reactions seem to go entirely to completion, just as the equation is written, whereas other reactions seem to go only part of the way.[1] The reactions you will study in today's experiment are of the second, more common, type.

Lots of reactions proceed around 99% of the way to completion and many go less than 1% of the way. Very few go around 50%. That is why it is so easy to fall into the trap of believing that they go all of the way or not at all. When ammonia gas is dissolved in water, it reacts with the solvent to form some ammonium ions, NH_4^+, and an equal number of hydroxide ions, OH^-.

$$NH_3 + H_2O \rightleftharpoons NH_4^+ + OH^-$$

This reaction is a prime example of one which proceeds only slightly (i.e. relatively few of the product ions are formed). In dilute solution at room temperature, the reaction proceeds only 1% or 2% of the way (see figure 10.1). On the other hand, the reaction of the acid hydrogen sulfide, H_2S (sometimes called rotten egg gas), with sodium hydroxide, NaOH, in solution proceeds nearly to completion according to the equation

$$H_2S + 2\,NaOH \rightleftharpoons Na_2S + 2\,H_2O.$$

---

[1]The combustion reaction of methane gas, CH_4, is one which might be considered to proceed entirely to completion, at least insofar as can be determined experimentally.

$$CH_4 + 2\,O_2 \rightarrow CO_2 + 2\,H_2O$$

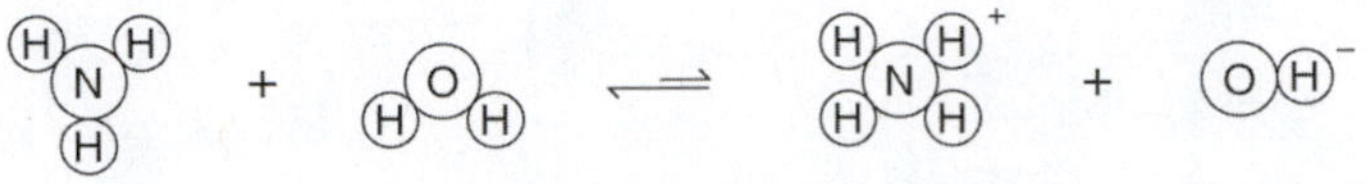

▲ **FIGURE 10.1** At equilibrium, molecules of NH_3 and H_2O combine to form a few NH_4^+ ions and OH^- ions, and these ions then recombine to re-form the original reactants. Both processes occur simultaneously. This equilibrium favors the left side of the reaction (i.e., most of the material exists as NH_3 and H_2O molecules). Note the reverse arrow is longer.

Reversibility and Equilibrium

A great many reactions stop short of completion because the products that are made react with each other to re-form the reactants; that is, a reaction (however slight) occurs in the reverse direction. Therefore, we say the reaction is *reversible*.[2] After a reaction is initiated, its speed begins to slow as the reactants become depleted. (As the reactant population thins out they have a harder time finding each other to react.) As the products are accumulated, their reaction, in the reverse direction, begins to accelerate. Finally, the reaction in the reverse direction attains the same rate as the forward reaction, and the system undergoes no further **net** change in concentration. The reaction may appear to have ceased, but actually both forward and reverse processes are occurring at the same rate at the same time. The reaction is, therefore, said to be in a state of dynamic equilibrium, or *at equilibrium* (figure 10.1).

Attainment of equilibrium in a reaction may happen very quickly (microseconds) or it may take days, but when a reaction has reached equilibrium (settles down), reactants and products are all present to some extent. Ammonia solution,[3] in the first example, has properties which suggest that all four species are present: ammonia, NH_3 (pungent odor); water, H_2O (wetness); ammonium ions, NH_4^+ (electrical conductivity); and hydroxide ions, OH^- (effect on litmus).

Shifting Equilibria

Once chemical equilibrium is established, it can be affected (disturbed) in such a way that it is momentarily upset. It then becomes established again with either a higher proportion of products (if it is shifted toward the right) or a higher proportion of reactants (if it is shifted toward the left). The equilibrium shift is an attempt to reestablish equilibrium after a disturbance such as the addition of a reactant, the removal of a product, or a change in the temperature or pressure. Since the speed of the forward reaction is governed by the concentration of the reactants on the left, adding a reactant would, at least momentarily, cause the forward reaction to go faster than the reverse reaction, thus effecting an equilibrium shift to the right (toward completion). Similarly, adding products or removing reactants would cause a shift in the reverse direction (to the left).

Law of Chemical Equilibrium—
A Theoretical Explanation for Equilibrium (Optional)

It is possible to grasp the basic concept of equilibrium without fully understanding the law that governs the process. The following is presented so students may gain some appreciation of the order and predictability (even beauty) that results when inanimate ions and molecules randomly bump into each other and react.

The concentrations of all ions and molecules at equilibrium are known to behave according to a simply stated equilibrium law. Applied to the reaction of ammonia with water,

$$NH_3 + H_2O \rightleftharpoons NH_4^+ + OH^-,$$

---

[2]The reverse arrows in the equations indicate that both forward and reverse reactions occur simultaneously. When one arrow is longer than the other, it indicates the direction in which the reaction has the strongest tendency to proceed.

[3]Often called ammonium hydroxide.

the law is as follows: the product of the concentrations of the substances formed, divided by the product of the concentrations of the reactants, is always a constant at a given temperature. This relationship, with its constant, is written

$$\frac{[NH_4^+][OH^-]}{[NH_3]} = K = 1.8 \times 10^{-5} \text{ at } 25°C$$

where $[NH_4^+]$ represents the concentration of ammonium ion in moles per liter, and $[OH^-]$ and $[NH_3]$ represent the molar concentrations of those species. Note that the species on the right (products) are in the numerator and the species on the left (reactants) are in the denominator.[4] The equilibrium relationship can tell us how an equilibrium will shift in response to a disturbance (a change in concentration of one of the species). If, for example, the numerator becomes larger, the concentrations in the denominator must also increase so that the equilibrium law is obeyed (i.e., so that the fraction remains equal to the constant). A large equilibrium constant (a large numerator) tells us the reaction goes nearly to completion (to the right) as written. A small constant indicates that the reaction proceeds only slightly to the right before equilibrium is reached.

The Reverse Reaction

Equilibrium may be approached from either side; the direction of approach makes no difference in the final equilibrium point. For any reaction, the equilibrium point can be reached either by allowing the reaction to proceed in the forward direction as written or by mixing an equivalent quantity of the end products and allowing the reaction to proceed in the reverse direction. This can be illustrated by one of the most important types of equilibrium, acid-base equilibrium.

Consider the acid-base reaction between H_2S and NaOH cited earlier and written here in ionic form as

$$H_2S + 2 Na^+ + 2 OH^- \rightleftharpoons 2 Na^+ + S^{2-} + 2 H_2O$$

or in net ionic form (including in the equation only those species involved) as

$$H_2S + 2 OH^- \rightleftharpoons S^{2-} + 2 H_2O.$$

When equivalent amounts of the acid, H_2S, and the base, NaOH, are mixed together (1 mole of acid to 2 moles of base), the reaction proceeds nearly to completion, but a little H_2S and OH^- remain unreacted. The resulting solution therefore has excess OH^- ions, as revealed by a litmus-paper test. When equivalent quantities of sodium sulfide, Na_2S, and water are mixed together, the reaction (which is the reverse of the previous reaction) proceeds only slightly, but enough to show the presence of OH^- ions.

$$S^{2-} + 2 H_2O \rightleftharpoons H_2S + 2 OH^-$$

This is why the final solution from the reverse reaction is indistinguishable from the final solution obtained from the original reactants.

Hydrolysis means reaction with water. In the foregoing example, S^{2-} (salt) ions react with water to form H_2S (an acid) and OH^- (a base). Hydrolysis of a salt is the exact reverse of neutralization. Typically,

---

[4]The concentration of water in a dilute solution is essentially constant at 55.5 moles/L, so it is incorporated into K and does not appear in the expression to the left.

the reaction of acid with base (neutralization) is predominant, but the reverse of that reaction (hydrolysis) is also important enough to be noticeable. When the acid and base involved are both strong (highly ionized), the equilibrium lies far to the right. However, when the reaction involves weak (slightly ionized) acids or weak bases, it is not nearly as complete.

In this experiment you will perform a number of tests on some of the reactions we have discussed, tests designed to see if the reactions are at equilibrium and whether the equilibria can be shifted to the right or to the left. Your main task will be to use the ideas presented in this discussion to explain what you observe.

Pre-Laboratory Questions | 10

1. What does the term *equilibrium* mean in everyday life? Is a child's seesaw in a state of equilibrium when it is at rest in the horizontal position?

2. What is meant by a reversible chemical reaction? Does a reversible reaction come to equilibrium?

3. In a reaction at equilibrium, why does changing the concentration of a reactant shift the equilibrium?

4. What is meant by the neutralization of an acid with a base? What is hydrolysis? How is hydrolysis related to neutralization?

EXPERIMENT

1. Chromate-Dichromate Equilibrium

Yellow chromate ions, CrO_4^{2-}, react with acid, H^+, to form orange dichromate ions, $Cr_2O_7^{2-}$, according to the equation

$$2\,CrO_4^{2-} \; + \; 2\,H^+ \; \rightleftharpoons \; Cr_2O_7^{2-} \; + \; H_2O.$$
$$\text{(Yellow)} \qquad\qquad\qquad \text{(Orange)}$$

Into 2 mL of 0.3 M K_2CrO_4 mix a couple of drops of 2 M H_2SO_4 (a source of H^+ ions) until you see a color change. Carefully add a few drops of 2 M NaOH (a source of OH^- ions) until the color changes again. Repeat the addition of acid until the color again changes. Explain the observed color changes by pointing out shifts in the equilibrium. You should be aware that H^+ (acid) and OH^- (base) remove each other from solution to form H_2O, thus accounting for the shifts, right or left. *Note:* be careful to dispose of the chromate compounds properly. (Answer Question 1, a–f on the Observations and Results sheet.)

Dispose of waste in the appropriate container.

2. Equilibria Involving Ammonia

This series of tests is designed to identify some species present in an ammonia, NH_3, solution.

A. Detection of Ammonia, NH_3

An ammonia, NH_3, solution (ammonium hydroxide) often is used in testing for the presence of copper(II) ions, Cu^{2+}. The reaction of these two species in solution produces a deep blue complex ion, $Cu(NH_3)_4^{2+}$.

$$Cu^{2+} \; + \; 4\,NH_3 \; \rightleftharpoons \; Cu(NH_3)_4^{2+}$$
$$\text{(Deep blue)}$$

Conversely, copper(II) ions can be used as a test for the presence of ammonia, NH_3. To 1 mL of 0.1 M copper(II), $Cu(NO_3)_2$, solution, add just enough drops of 1 M NH_4OH (NH_3 solution) to make a clear, deep blue solution. Is NH_3 present in the solution labeled ammonium hydroxide?

Now suggest a way to shift the equilibrium (involving the complex ion) back to the left, and carry it out. [Hint: Add something that will effectively remove (destroy) the base, ammonia.] See if you can make the reaction go forward and backward by alternately adding small amounts of H^+ and NH_3 solutions. (Answer Question 2A, 1–4.)

B. Evidence for the Presence of Hydroxide Ion, OH^-

Place 1 mL of 2 M NH_4OH in a test tube. Now test the ammonia solution for the presence of OH^- ion by using red litmus paper or other pH paper. Red litmus or pH paper turns blue in a water solution of a base. Blue litmus paper turns red in an acidic solution.

Add 1 mL of 1 M magnesium chloride, $MgCl_2$, to the ammonia solution and observe the result. Magnesium hydroxide, $Mg(OH)_2$, is an insoluble white precipitate which forms when sufficient Mg^{2+} ions and OH^- ions are both present in solution. Write the equation for the reaction. Are OH^- ions present in ammonia solutions? Recall that the equation for the equilibrium of ammonia solution is

$$NH_3 + H_2O \rightleftharpoons NH_4^+ + OH^-.$$

(Answer Question 2B, 1–2.)

C. Approaching the Equilibrium from the Product Size

Now see if you can achieve the same equilibrium conditions by starting with the substances on the right side of the equation, NH_4^+ and OH^-. To 1 mL of 2 M sodium hydroxide, NaOH, add about 0.2 g of solid ammonium chloride, NH_4Cl, and stir until all is dissolved. Remember that each ion in an ionic compound behaves independently of the others. You have Na^+, NH_4^+, OH^-, and Cl^- ions present. Will any of these react with each other? Test for the presence of ammonia, NH_3, first by smelling and then by adding 0.5 mL of 0.1 M $Cu(NO_3)_2$ solution (Cu^{2+} ions). Make an additional test. Place a strip of dry and a strip of moist red litmus or pH paper in the mouth of the test tube,[5] heat gently, and note what happens. To explain your observations, think of what occurs on the wet paper and why the wet one gives a definite test. (Answer Question 2C, 1–5.)

3. Hydrolysis Equilibria

A. Hydrolysis of Bismuth Chloride

Dissolve about 0.3 g of bismuth chloride, $BiCl_3$, in 1 mL of water in a test tube and note the result. [Bismuth chloride, $BiCl_3$, partially hydrolyzes (reacts with water) to form bismuth oxychloride, BiOCl.] Test the solution for acidic or basic properties with litmus or pH papers. (Answer Question 3A, 1–4.)

B. Hydrolysis of Sodium Sulfide

Dissolve about 0.5 g of sodium sulfide, Na_2S, in one or two milliliters of water and note any effect, including odor. Then test the solution with litmus paper. (Answer Question 3B, 1–3.)

Dispose of all waste from parts 1-3 of this experiment in the appropriate containers.

---

[5]First rinse the mouth of the test tube with distilled water. Remember that you poured in some base, NaOH. What would that do to the litmus paper?

QUESTIONS AND PROBLEMS

1. Write equations for (a) the hydrolysis of $TiCl_4$ and (b) the neutralization of $Ba(OH)_2$ with H_2SO_4.

2. Iron(III) ion, Fe^{3+}, reacts with thiocyanate ion, SCN^-, to form the blood-red complex ion $Fe(SCN)^{2+}$

$$Fe^{3+} + SCN^- \rightleftharpoons \underset{\text{(Blood-red)}}{Fe(SCN)^{2+}}.$$

Tell whether the intensity of the red color of the solution would increase or decrease for each of these changes: (a) addition of more Fe^{3+} ion, (b) addition of more SCN^- ion, (c) addition of more $Fe(SCN)^{2+}$ ion, (d) dilution with water, (e) addition of a catalyst, and (f) addition of a base to precipitate $Fe(OH)_3$.

3. The esterification reaction

$$C_2H_5OH + CH_3COOH \rightleftharpoons CH_3COOC_2H_5 + H_2O$$

has the following equilibrium relationship at 25°C in an organic solvent.

$$\frac{[CH_3COOC_2H_5][H_2O]}{[C_2H_5OH][CH_3COOH]} = 4.0$$

a. What concentration of water would be present in an equilibrium mixture in which the concentrations of the other reactants are all 2 M?

b. Would equilibrium exist if $CH_3COOC_2H_5$ and H_2O were 2 M each and C_2H_5OH and CH_3COOH were 1 M each?

c. Describe what would occur if one were to mix C_2H_5OH and CH_3COOH together, each initially at 3 M concentrations.

d. In which directions would a shift occur if the initial concentrations of all four species were 1 M?

EXPERIMENT

Chemical Reactions Can Go Forward and Backward

10

OBSERVATIONS AND RESULTS

1. Chromate-Dichromate Equilibrium

a. Write the equation for the chromate-dichromate equilibrium.

b. What color change did you observe when H_2SO_4 was added? Explain how the addition of H_2SO_4 caused the color of the solution to change.

c. Was there a shift in the equilibrium? In which direction?

d. Considering the discussion of what causes an equilibrium to shift, and noting the equation, would you say H^+ or SO_4^{2-} was responsible for the change in color? Explain.

e. What did you observe, and in which direction did the equilibrium shift when NaOH was added? (Remember OH^- ions react with and remove H^+ ions.)

f. How did OH^- ion cause the color of the solution to change?

2. Equilibria Involving Ammonia

A. *Detection of Ammonia,* NH_3

    1. Describe what happened when you added ammonia solution to the $Cu(NO_3)_2$ solution.

    2. Do the copper(II) ions indicate that ammonia, NH_3, is present in the ammonia solution? Explain.

    3. What did you suggest as a way to shift the equilibrium (involving the complex ion) to the left?

    4. Did it work? If so, explain how it worked.

B. *Evidence for the Presence of Hydroxide Ion,* OH^-

    1. What is the evidence for the presence of OH^- ions in the ammonia solution? (Give two indications.)

    2. Complete and balance the net ionic equation for the reaction of $MgCl_2$ with NH_4OH. (Leave out the Cl^- spectator ions.)

$$Mg^{2+} + NH_4OH \rightleftharpoons Mg(OH)_2 +$$

C. *Approaching the Equilibrium from the Product Side*

    1. When ammonium chloride, NH_4Cl, and sodium hydroxide, NaOH, solution were mixed, which ions actually reacted with each other?

2. Complete the following equations that describe the reactions that occurred.

$$NH_4^+ \quad + \qquad \rightleftharpoons \quad NH_3 \quad +$$

$$Cu^{2+} \quad + \qquad \rightleftharpoons$$

3. Did copper(II) ion, Cu^{2+}, confirm the formation of ammonia? What was your evidence?

4. Did dry litmus change color in the vapor? Did wet litmus? Describe the color change if there was one.

5. What ion was formed when NH_3 came in contact with wet litmus? How do the presence of this and the change in color of the litmus paper prove that ammonia gas was escaping from the solution?

3. Hydrolysis Equilibria

A. Hydrolysis of Bismuth Chloride

1. Describe what happened when bismuth chloride, $BiCl_3$, was placed in water.

2. On the basis of litmus tests, was the solution acidic or basic?

3. Complete and balance the equation for the hydrolysis of bismuth chloride.

$$BiCl_3 \quad + \quad H_2O \quad \rightleftharpoons \quad \underset{\substack{\text{Bismuth} \\ \text{oxychloride}}}{BiOCl} \quad +$$

4. How does the equation account for the fact that the solution became acidic or basic?

B. *Hydrolysis of Sodium Sulfide*

1. Complete the net ionic equation for the reaction that occurs when sodium sulfide, Na_2S, hydrolyzes in water.

$$S^{2-} \;+\; 2\,H_2O \;\rightleftharpoons\; \underline{\hspace{2cm}} \;+\; \underline{\hspace{2cm}}$$

2. What was the evidence that Na_2S hydrolyzed in water?

3. Was the solution acidic or basic?

Weights of Reactants and Products: Preparation of a Metal | 11

OBJECTIVES

1. To relate calculations from formulas and equations to observed results.
2. To gain experience in the collection of quantitative data.
3. To learn some steps in metal production by doing a typical preparation.
4. To illustrate the concept of yield.

DISCUSSION

Chemical formulas and equations provide a wealth of quantitative information about compounds and reactions. Proficiency in dealing with chemical quantities is one of the major skills of a person educated in the sciences. In this experiment you will perform a metallurgical operation, synthesizing lead from a pure ore, lead(II) carbonate. To accomplish this you will first heat lead(II) carbonate, $PbCO_3$, to produce its oxide, PbO. You will then heat the lead(II) oxide with carbon to produce lead metal. You will make a quantitative study of the process to determine the weights of the lead(II) carbonate, lead(II) oxide, and lead metal and compare the measured (experimental) weights of the lead(II) oxide and lead with the calculated or theoretical amounts.

Information in a Formula

Recall what information a chemical formula contains. Since it gives the number of atoms of each kind in the molecule, it reveals the composition of the compound by weight as well as by number of atoms.

For example, the formula for phosphorus pentachloride, PCl_5, tells us that a molecule of PCl_5 has one phosphorus atom and five chlorine atoms (a 1:5 ratio in the compound). Since we know that the atomic weight of phosphorus is 31 and that of chlorine is 35.5, the molecular weight of PCl_5 is $31 + 5(35.5) = 31 + 177.5 = 208.5$. A mole of PCl_5, therefore, weighs 208.5 g. Of that amount, 31 g is contributed by phosphorus and 177.5 g by chlorine.

It is a simple matter to calculate the percentage (by weight) of each element in a compound. For example, for PCl_5:

Percentage of phosphorus: $\dfrac{31}{208.5} \times 100 = 14.9\%$

Percentage of chlorine: $\dfrac{177.5}{208.5} \times 100 = 85.1\%$

We can also easily calculate the weight of any element in a given quantity of a pure compound. The weight of chlorine in a 5-gram sample of PCl_5 can be calculated as follows:

Weight of Cl in 5 grams of PCl_5 is the fraction[1] of Cl in PCl_5 times the

weight of the PCl_5 sample $= 0.851 \times 5$ g $= 4.26$ g.

How are the formulas of compounds determined? What we have said about the calculations of weights from formulas can be reversed to determine a formula. For example, if the weights of the elements in a compound are determined by chemical analysis, these weights along with the atomic weights can be used to find the simplest or empirical formula experimentally.

Information in a Chemical Equation

A chemical equation tells a great deal about a reaction. In addition to identifying the reactants and products the balanced equation provides information concerning the quantities of each reactant and product. For example, the equation $P_4 + 10\,Cl_2 \rightarrow 4\,PCl_5$, for the formation of PCl_5 from its elements, says

1. One molecule of P_4 combines with 10 molecules of Cl_2, to form 4 molecules of PCl_5.

2. One mole of P_4 (or 4 moles of phosphorus atoms, 124 g) combines with 10 moles of Cl_2 (or 20 moles of chlorine atoms, 710 g) to form 4 moles of PCl_5 (834 g).

These quantities of phosphorus and chlorine may be called the standard batch (or recipe) for producing the indicated amount of PCl_5. Multiples of these amounts, whether larger or smaller, produce an amount of phosphorus pentachloride proportionately larger or smaller. It is the ratio of moles and ratio of grams that is significant. It is not difficult to make a calculation from a chemical equation. Use the following steps to calculate the weight of Cl_2 required to react with 3 grams of P_4 to form PCl_5 and to determine the weight of the resulting product.

1. We are given 3 g of P_4 which is $\dfrac{3}{124} = 0.024$ mole.

2. Calculate the moles and grams of Cl_2 and PCl_5 using mole ratios (coefficients) from the equation:

for Cl_2: $\quad 0.024$ mole $\times \dfrac{10}{1} = 0.24$ mole; $0.24 \times 71 = 17$ g of Cl_2

for PCl_5: $\quad 0.024$ mole $\times \dfrac{4}{1} = 0.096$ mole; $0.096 \times 208.5 = 20$ g of PCl_5

---

[1]The fraction is the part of the whole (1), whereas the percentage is the part of 100.

Minerals, Ores, and Metallurgy

An ore is a deposit of a valuable mineral usually mixed with unwanted minerals. Table 11.1 gives the mineral sources of some important metals. You will note that minerals are compounds. Metals generally occur in nature in the form of cations, which are chemically changed to neutral atoms in metallurgy.

Isolating a Metal from an Ore

The method of isolating a metal from its mineral compound in the ore depends largely on the activity of the particular metal. The most active metals are so stable in compounds that they cannot be chemically separated under ordinary conditions. These active metals such as sodium, magnesium, and aluminum are obtained by sending an electric current through the molten material.

Some of the cations of the least active metals are so weakly bound to the nonmetal ions that heating decomposes them and a few, like gold, are not even combined but exist in the free metallic state. An example of a weakly held metal is mercury in mercury sulfide, HgS. Experienced prospectors test rocks in the field for cinnabar by heating them to see if metallic mercury is liberated (at risk to their health). Most metals lie between these extremes, and are extracted by treatment with a reducing agent such as carbon or hydrogen. For example, the steel industry uses carbon to produce iron from hematite, Fe_2O_3, in the blast furnace according to the equation $2\,Fe_2O_3 + 3\,C \rightarrow 4\,Fe + 3\,CO_2$.

There are several distinct steps in typical metallurgical processes.

1. *Concentration.* The crude ore is separated or made more concentrated by several methods, such as flotation, decomposition, or screening.

2. *Roasting.* Many ores, particularly sulfides, are converted to the oxides by heating in air:

$$2\,PbS + 3\,O_2 \rightarrow 2\,PbO + 2\,SO_2.$$

3. *Reduction.* Carbon or another reagent reduces the ore to the metal.

4. *Refining.* Impurities are removed from the metal by several means.

TABLE 11.1 Metals and Minerals

Metal	Mineral	Formula
Aluminum	Bauxite	$Al_2O_3 \cdot xH_2O$
Copper	Chalcocite	Cu_2S
Iron	Hematite	Fe_2O_3
Silver	Hornblend	$AgCl$
Zinc	Sphalerite	ZnS
Lead	Galena	PbS
	Cerussite	$PbCO_3$
Titanium	Rutile	TiO_2
Mercury	Cinnabar	HgS

Lead from Cerussite

In today's experiment, you will simulate the production of metallic lead from the mineral cerussite, which is lead(II) carbonate, $PbCO_3$. You will decompose $PbCO_3$ to the more reactive lead(II) oxide, PbO, by roasting (heating).

$$PbCO_3 \rightarrow PbO + CO_2$$

The lead(II) oxide, in turn, will be reduced to lead metal by heating with carbon.

$$2\,PbO + C \rightarrow 2\,Pb + CO_2$$

ENVIRONMENT, CULTURE, AND CHEMISTRY
The Dilemma of the Metal Supply

Those shiny, necessary materials we call metals are both a bane and a blessing. Modern society depends on metals; in some situations they are irreplaceable. However, care must be taken with some of the heavy metals, such as lead and mercury, because of their toxicity. Lead is known to affect human function and development, particularly of children. A common source of lead poisoning is the paint on old buildings. Mercury in the amalgam used to fill cavities in teeth is another area of concern. On the other hand, the ubiquitous sodium, potassium, calcium, and magnesium (as ions) are both abundant and physiologically essential to the body. Some metals are so inert, like gold, that they are harmless and decorative. Tantalum, which is compatible with body tissue and fluids, is used for bone parts replacement. Supplying our needs for metals is a huge environmental challenge that often provokes controversy at many stages: mining, concentrating, smelting, extruding, machining, processing, recycling, and disposing. The problems include the pollution of air, streams, and landscape; the threatening of species and ecosystems; and the safety of workers. Our most accessible, richest deposits are becoming depleted, and we have been forced to resort to leaner sources. Today, as with petroleum, the search for metals has turned from land to the sea floor. In the twentieth century, rich metal deposits were discovered on the ocean floor, usually where the continental plates join, and sea water can seep deep into the earth's hot magma. The metals containing minerals dissolve in the hot solution, and rise to mix with the cold ocean waters, where they crystallize and form rich deposits. These warm plumes also support unexpected ecosystems where thousands of new species, particularly microbes, have been discovered. Obtaining metals from this unusual source presents a unique environmental problem, similar to ordinary mining. Even though fragile, nature often shows a remarkable ability to adapt, especially if managed responsibly. For example, it was observed that the area around the mineral-rich ocean vent holes, following a volcanic eruption in the East Pacific in 1991 that obliterated the sea life on the ocean floor, showed a surprising recolonization within just two years. This observation suggests that with good management, it may be possible to keep the effects of mining activity from exceeding the rate of recovery of these habitats. Ultimately, as with all environmental problems, reasonableness can be the key to success.

Pre-Laboratory Questions | 11

1. How many atoms are present in each molecule of phosphorus oxide, P_4O_6?

2. Calculate the weight percentages of phosphorus and oxygen in P_4O_6.

3. If the ore in today's experiment is pure cerussite, $PbCO_3$, (a) how many moles of $PbCO_3$ are there in 7.5 g of the starting material? (b) how many moles of PbO would form when the $PbCO_3$ is heated? (c) how many moles of Pb metal would the heated ore form? (d) how many moles of carbon are required to reduce the PbO to Pb according to the equation given? (e) is an excess of charcoal used in the experiment? How much of an excess?

4. Write equations for the reactions that might be used to make zinc metal from a common zinc ore such as ZnS.

5. What would happen in today's experiment if a hot blue flame rather than a cooler yellow one were used in heating the crucible containing $PbCO_3$? What error would be caused if the crucible were underheated?

EXPERIMENT

1. Decomposition of Lead(II) Carbonate

Suggestion: If possible, the parts of this experiment that involve heating should be done under an exhaust hood.

Your goal in this experiment is to obtain quantitative results. Make all weight measurements carefully and to the maximum accuracy of your balance.

Zero (tare) the weight of a clean, dry crucible (without a lid). Then add about 2.0 g of lead(II) carbonate and record its weight. (Answer Question 1a on the Observations and Results sheet.) Heat the crucible and lid with a *cooler, yellow flame* for 30 minutes.[2] The crucible should be 4 to 5 cm above the top of the burner so that the flames extend up the sides of the crucible (figure 11.1). A black carbon deposit may form on the sides of the crucible but it can be wiped off easily. Stir the material frequently with your spatula so that all of it is exposed to the hottest part (bottom) of the crucible, taking care not to transfer any solid from the crucible. The color of the material in the crucible should change from white to yellowish orange (the color of lead(II) oxide, PbO) during the heating process. Place the crucible on a hot pad to cool and then determine its weight to the maximum accuracy. It should weigh less than before heating because lead(II) carbonate decomposes to lead(II) oxide and carbon dioxide is expelled. Record the weight of lead(II) oxide (answer Question 1b) and perform the calculations (answer Question 1, c–d) concerned with the conversion of lead(II) carbonate and lead(II) oxide. (Answer Question 1, f–g.)

2. Reduction with Carbon

When you have weighed the crucible, add approximately 1 g of powdered wood charcoal (carbon) to the crucible and mix it in thoroughly with a spatula. Replace the lid on the crucible and heat it for 15 minutes with the hottest Bunsen flame that you can obtain. (The center cone of the flame should just reach the bottom of the crucible.) During the last few minutes, remove the lid and stir the contents with your spatula. A pool of shiny lead should be visible. If no lead is visible, continue to heat and if necessary, stir in another portion of charcoal (about 0.5 g).

When a pool of lead is visible, let the crucible cool on the hot pad. Next, pour the contents of the crucible into a 250-mL beaker half filled with water. Stir the mixture; the lead will settle to the bottom and most of the carbon will float. Decant the water and carbon, leaving the lead in the bottom of the beaker. Repeat the washing process several times, until the lead is reasonably free of carbon.

Dispose of waste in the appropriate container.

At this point, if there are one or a few large globules of lead, remove them, dry with a paper towel, and weigh the metal. If the lead is present as many tiny granules, however, decant all of the water and transfer the lead back into your crucible with a spatula. Heat the crucible at full heat. The lead will melt, enabling you to push all of the lead droplets into a large globule with your spatula. Remove the crucible from the burner with the forceps, and pour the molten lead into a beaker of water. Remove the lead, dry, and weigh it. Record the weight of the lead (answer Question 2a). Note its physical properties. Hit it with a hammer or heavy tool. Place the lead in the appropriate container. Perform the calculations to determine the percentage of lead recovered (answer Question 2, b–c). This is the yield that can approach but not exceed 100%. (Answer Question 2, d–e.)

---

[2]The hotter, blue flame fuses lead(II) oxide which, on cooling, is a hard layer that adheres to the crucible.

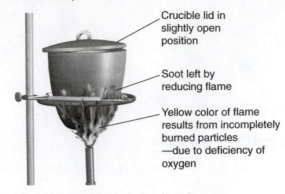

Crucible lid in
slightly open
position

Soot left by
reducing flame

Yellow color of flame
results from incompletely
burned particles
—due to deficiency of
oxygen

▲ **FIGURE 11.1** A crucible positioned in a cool (reducing) yellow flame.

QUESTIONS AND PROBLEMS

1. A lead ore contains 14.7% lead. Assuming 95% recovery of the lead, how much ore must be processed to make a ton of lead?

2. If the pure minerals were used, which ore—galena or cerussite—would be the richer source of lead? Show calculations indicating the percentages of lead in each ore.

3. Complete and balance the following equations, which illustrate the procedure for recovering lead from galena:

    a. Roasting: $PbS + O_2 \rightarrow$

    b. Reduction: $PbO + H_2 \rightarrow$

4. Tetraethyl lead, $Pb(C_2H_5)_4$, was once used in regular gasoline to improve its burning characteristics but has since been banned because of health concerns. If 0.4 g of lead were present in each gallon of gasoline, calculate the grams of lead released into the environment each year by each driver who consumed three gallons per day of this gasoline.

5. Name some contemporary uses for lead and some former uses that have been discontinued for environmental reasons.

EXPERIMENT

Weights of Reactants and Products: Preparation of a Metal

11

OBSERVATIONS AND RESULTS

1. Decomposition of Lead(II) Carbonate (Cerussite)

a. Weight of $PbCO_3$ _____ g

b. Weight of PbO obtained _____ g

c. Moles of PbO obtained (show calculation) _____ mole

d. Moles of PbO expected based on calculation from weight of $PbCO_3$ (show calculation) _____ mole

e. Percentage error $= \dfrac{\text{(Difference between moles of PbO obtained and moles expected)}}{\text{Moles of PbO expected}} \times 100$

$= $ _____ %

f. Is your weight of PbO higher or lower than expected?

g. Offer a possible explanation for the difference from the expected value.

2. Reduction of Lead(II) Oxide to Lead

   a. Weight of your recovered lead metal   _____ g

   b. Weight of lead present in your lead(II) carbonate based
      on the percentage of lead in lead(II) carbonate (use formula
      to calculate the percentage; show calculation)   _____ g

   c. Percentage of lead recovered (yield) $= \dfrac{(\text{Weight recovered lead}) \times 100}{\text{Weight lead in lead(II) carbonate}}$

       $= $ _____ %

   d. Describe several physical properties of lead metal.

   e. Note one difference in physical properties between lead carbonate and lead oxide.

The pH Scale and the Use of Indicators | 12

OBJECTIVES

1. To become familiar with acids and bases in water solution.
2. To master the pH scale and pH measurement.
3. To learn the use of acid-base indicators.
4. To become aware of the degree of acidity or basicity (pH) of some common materials.

DISCUSSION

Acids are commonly described as sour tasting, like vinegar or lemon juice. Water solutions of bases are said to taste bitter and feel slippery, but most strong bases like lye (sodium hydroxide) are too bitter and harsh to be tasted. Examples of milder bases are slaked lime (calcium hydroxide) and ammonium hydroxide.[1] Acids are characterized by the presence of H^+ ions (actually H_3O^+ ions) in water solutions, and bases are typically those compounds having OH^- ions. The H^+ ions of acids and the OH^- ions of bases react with each other to form water, so acids and bases are said to neutralize each other. For this reason they are considered to be opposites. In this experiment you will learn to measure and to express in concise terms the degree of acidity or basicity of solutions. The famed pH scale, so casually used by television advertisers, was devised precisely for this purpose.

Ionization of Water

An understanding of acidity and basicity in water solution (most of the important acid-base chemistry takes place in water) rests on the concept of the ionization of water. Pure water ionizes only about 0.0000002%; the ionization is shown in the following equation:

$$H_2O \;+\; H_2O \;\rightleftharpoons\; H_3O^+ \;+\; OH^-$$

For convenience, the hydronium ion, H_3O^+, is often written as H^+; then the ionization of water can be simplified to

$$H_2O \;\rightleftharpoons\; H^+ \;+\; OH^-.$$

---

[1] Solutions of ammonia, NH_3, in water are often called *ammonium hydroxide*, NH_4OH, because ammonium ions, NH_4^+, and hydroxide ions, OH^-, are formed in low concentration in the reaction with water.

$$NH_3 \;+\; H_2O \;\rightleftharpoons\; NH_4^+ \;+\; OH^-$$

You can see from the equation that in pure water there are as many H^+ ions as there are OH^- ions; that is, the H^+ ion concentration, $[H^+]$, and the OH^- concentration, $[OH^-]$, are equal. The actual concentration of each is minute. In pure water these are both known experimentally to be only 1×10^{-7} (0.0000001) mole/L, or 1×10^{-7} molar (*M*).

$$[H^+]=[OH^-]=1\times10^{-7}\ \text{mole/L}$$

In acid solution the H^+ ion concentration is greater than 10^{-7} mole/L and the OH^- ion concentration is less than that: $[H^+]>[OH^-]$. In basic solution the opposite is true and OH^- ions predominate: $[OH^-]>[H^+]$. The dissociation of water can be described by an equilibrium expression, just like any other reaction at equilibrium. The equilibrium constant for water, K_w, is found by multiplying the concentration of H^+ and OH^- ions in pure water, 1×10^{-7} mole/L for each.

$$K_w=[H^+][OH^-]=(1\times10^{-7})^2=1\times10^{-14}$$

In any aqueous solution at room temperature, the product of the H^+ and OH^- ion concentrations is always equal to 10^{-14}. As the concentration of H^+ ions becomes greater (in acidic solutions), the OH^- ion concentration necessarily becomes smaller; the product of the two always equals 10^{-14}. Both kinds of ions are present in all water solutions. The acidity or basicity of the solution depends on which ion predominates. If their concentrations are equal (at 10^{-7} mole/L), the solution is neutral. For example, if the H^+ ion concentration is 10^{-3} mole/L, then the OH^- ion concentration can be found from the equation for the equilibrium constant for water. Since $[H^+][OH^-]=10^{-14}$,

$$10^{-3}\times[OH^-]=10^{-14},$$

$$[OH^-]=10^{-11}\ \text{mole/L}.$$

The pH Scale

Based on your understanding of H^+ and OH^- ion concentrations, you are now ready to examine the meaning of the pH scale. This scale constitutes a highly convenient method for specifying the acidity (or basicity) of a solution. The pH is actually the exponent of the H^+ ion concentration (the power to which 10 is raised) with its sign changed.[2] A solution with an H^+ ion concentration of 10^{-3} mole/L has a pH of 3. A pH of 6 means an H^+ ion concentration of 10^{-6} mole/L, which also means an OH^- ion concentration of 10^{-8} mole/L. When any one of the quantities pH, $[H^+]$, or $[OH^-]$ is specified, the other two are also specified. Note the H^+ ion concentrations, OH^- ion concentrations, and corresponding pH values in table 12.1.

---

[2]pH is more elegantly defined mathematically as the logarithm of the reciprocal of the H^+ ion concentration,

$$pH=\log\frac{1}{[H^+]},\ \text{or as the negative logarithm of the } H^+ \text{ ion concentration, } pH=-\log H^+$$

TABLE 12.1 pH Values in Aqueous Solutions

$[H^+]$	$[OH^-]$	pH	
10^1	10^{-15}	-1	
$10^0 (1)$	10^{-14}	0	
10^{-1}	10^{-13}	1	
10^{-2}	10^{-12}	2	
10^{-3}	10^{-11}	3	Acidic
10^{-4}	10^{-10}	4	
10^{-5}	10^{-9}	5	
10^{-6}	10^{-8}	6	
10^{-7}	10^{-7}	7	Neutral
10^{-8}	10^{-6}	8	
10^{-9}	10^{-5}	9	
10^{-10}	10^{-4}	10	
10^{-11}	10^{-3}	11	
10^{-12}	10^{-2}	12	Basic
10^{-13}	10^{-1}	13	
10^{-14}	$10^0 (1)$	14	
10^{-15}	10^1	15	

All pH values less than 7 indicate acidic solutions, and all values greater than 7 indicate basic solutions. A neutral solution has a pH of 7 since the H^+ and OH^- ion concentrations are equal. It should also be emphasized that each pH unit on the scale means a tenfold increase or decrease in H^+ ion concentration from the previous number. Thus, a pH range of 0 to 14, for example, represents a range of concentration of H^+ (or OH^-) ions from 1 mole/L to 10^{-14} mole/L (from 1 to 0.00000000000001)—a tremendous range. Finally, the pH scale is not restricted to whole numbers. Calculation of fractional pH values requires the use of logarithms. Some fractional pH values and the corresponding concentrations of H^+ or OH^- ions are shown in table 12.2.

Indicators

Acid-base indicators are compounds which exhibit one color in an acid solution and a different color in a more basic solution. Not all indicators change color at the same pH. Some indicators change colors at pH 7, others at pH 4, pH 5, or pH 6, and some at pH 8, pH 9, or pH 10. The color shift of an indicator, from its "acidic color" to its "basic color," requires about 2 pH units, or 1 full pH unit on either side of the midpoint. Indicators are useful for telling us when solutions change from acidic to basic, and vice versa. In a titration where an acid and a base are gradually mixed together it is convenient to have an indicator present to signal when enough of the one has been added to neutralize the other. Refer to the appendix, table A.5, for a listing of common indicators, their color ranges, and their transition points.

TABLE 12.2 Some Fractional pH Values

[H$^+$]	[OH$^-$]	pH
2.0×10^{-3}	5.0×10^{-12}	2.70
7.7×10^{-5}	1.3×10^{-10}	4.11
5.0×10^{-9}	2.0×10^{-6}	8.30
3.0×10^{-12}	3.3×10^{-3}	11.52

Generally, indicators are available in laboratories as aqueous solutions. A few drops in a sample usually is sufficient to show a definite color. Sometimes indicators are impregnated in strips of paper that can be dipped into a solution. Litmus papers, which turn blue in base and red in acid, are common examples. Universal indicator papers, called pH paper, have become popular. They show a continuous change of color over a wide pH range, usually from very red for pH 1 to very blue for pH 14. Narrow-range pH papers, which can indicate by color hue the pH of a solution to within a few tenths of a pH unit, are also available.

Laboratories requiring rapid accurate pH measurements usually use a pH meter. The theory behind a pH meter is complicated, but it is easy to use: you place an appropriate electrode in a sample and read the pH on the instrument dial. The measurement is based upon the experimental fact that a change in hydrogen ion concentration can cause a change in the voltage of an electrochemical cell. The pH meter is fast and consequently is especially valuable when testing many samples. It is also preferred for dark or colored solutions where indicator colors may be obscured.

In today's experiment you will see how indicators work. You will explore their color ranges and locate the pH of their color transitions. You will do this by making some reference solutions of known pH and observing the colors of various indicators in each of them. You will then use your knowledge to determine the pH of several common materials. Finally, if time permits, you may have an opportunity to use the pH meter to check some pH values.

ENVIRONMENT, CULTURE, AND CHEMISTRY

Acid Rain, Erosion of Statues and Monuments, and Smog: Oxides of Carbon, Sulfur, and Nitrogen

A casual observer may be unaware of the effects of acids and other chemicals in nature because they often act so slowly and subtly. Carbon dioxide, CO_2, is the end product of food metabolism by humans and animals and from the rotting and burning of plant materials. In water, CO_2 becomes a weak acid, carbonic acid, H_2CO_3. Humans personally suffer no corrosive effects from carbonic acid because it is only slightly acidic, with a pH just below 7. However, carbonic acid is sufficiently acidic to dissolve limestone and marble, both calcium carbonate, $CaCO_3$, to form soluble calcium bicarbonate, $Ca(HCO_3)_2$. Therefore, limestone structures and statuary are gradually being eroded by carbonic acid. (Cement is also primarily limestone and undergoes the same erosion.) Occasionally, this erosion works to our enjoyment: visitors to caves, such as the spectacular Carlsbad Caverns, witness the products of this dissolving action that has occurred over eons of time in the forms of stalactites and stalagmites. Our modern industrial world, which depends on burning enormous quantities of coal and petroleum products, has greatly increased the production of CO_2.

Another destructive atmospheric gas is sulfur dioxide, SO_2. This gas is emitted in relatively small amounts from volcanic and other earth-venting action, but much greater quantities are formed from the oxidation of sulfur when burning coal, where it exists at low levels. Sulfur dioxide dissolves in water, making sulfurous acid, H_2SO_3, which in the presence of oxygen produces sulfuric acid, H_2SO_4. The oxides of nitrogen, a mixture of NO, NO_2, and other nitrogen oxides (called NO_x), are formed at the high temperatures in the interior of automobile engines when gasoline and oxygen combust. The brown color on the horizon around many cities is primarily caused by NO_2. When the nitrogen oxides come in contact with water they produce nitrous acid, HNO_2 and nitric acid, HNO_3.

All three of these gases are playing a very destructive role in the environment. SO_2 and NO_x, as gases, have a direct impact on human health, especially the health of children and adults with respiratory problems. Sulfur dioxide, SO_2 dissolved in moist air forms sulfurous acid and sulfuric acid, a mixture called acid rain. HNO_3 from NO_x also contributes to acid rain. Two of the most noticeable effects of acid rain are on forests and lakes, particularly at higher altitude: trees (mostly pines) initially develop a sickly yellow color, eventually dying, and fish have disappeared from mountain lakes and streams. Two of the principal causes of the death of fish are the increased acidity (drop in pH), and a significant elevation in the concentration of aluminum ion and other ions in the lake water, caused by the effect of acid on the surrounding soil. The disappearance of fish has had a direct impact on the abundance of certain birds, extending a long chain of environmental losses. Acid rain also has an effect at lower altitudes, most obviously on statues and buildings around the world. In Egypt, the Sphinx and the pyramids have suffered serious erosion and drastic steps are being taken to preserve them. Most Italian cities have been severely criticized for permitting erosion of some of their irreplaceable structures, such as Michelangelo's David and even the excavated ruins of ancient Pompeii. The statue of David, completed in 1504, was moved into a building in 1873. Currently it is surrounded with a "wall" of pure air to protect it. Finally, the nitrogen oxides, NO_x, play a major role in the formation of smog, a soup of NO_x, hydrocarbon combustion products, and other impurities caused by the interactions of the various components in the presence of oxygen. Smog is responsible for the hazy atmosphere and toxic air of many of the worlds' great cities. Several of those in the most-polluted top twelve are Beijing, Cairo, Rome, New Delhi, Vancouver, Tokyo, Melbourne, Paris, and Los Angeles. Los Angeles has improved significantly in the past few years because of tight control over engine emissions and strict burning laws.

Pre-Laboratory Questions | 12

1. Phenolphthalein is one of the most common indicators. What are its colors in acidic and in basic solutions?

2. What does pH mean? What is the pH of a neutral solution?

3. If a solution contains 0.01 mole of HCl in 10 L of solution, what is its molarity? Its H^+ ion concentration? Its pH?

4. What relationship exists between H^+ ion concentration and OH^- ion concentration? What is the OH^- concentration in an aqueous solution where the H^+ concentration is 10^{-4} M?

5. How many pH units are usually required for an indicator color transition?

6. What are the colors and the transition pH for methyl red, bromthymol blue, and alizarin yellow? (See the appendix, table A.4.)

EXPERIMENT

1. Color Characteristics of Some Indicators (a Class Activity)

In this part of the experiment you and your partner will prepare one of the sets of indicator solutions to be used by the entire class. The set will include 11 solutions in test tubes ranging in pH from 2 to 12, and it will show the colors given by a certain indicator over this pH range (see figure 12.1). Other groups will prepare similar sets with other indicators so you all can observe the colors given by various indicators. Recommended indicators are methyl orange (MO), phenolphthalein (Phth), bromthymol blue (BTB), alizarin yellow (AY), and methyl red (MR), but others may be substituted.

When this part of the experiment is finished there should be five different indicator reference sets labeled and ready for use. If the class is large enough there may be duplicate sets. Verify the actual colors of these sets with those listed in table A.5 in the appendix. If one of the sets is missing or improperly prepared, note this and use the correct colors shown in the table.

Prepare the 11 solutions as instructed below. Use clean test tubes and containers that have been *rinsed with distilled water* because small amounts of stray acid or base can cause large pH changes. The test tubes should be the same size and filled to the same depth (about two-thirds full). They should be placed in order from pH 2 to pH 12 in a test tube rack and labeled with the correct pH.

A.   Preparation of Solutions in the Acid Range, pH 2 to pH 6

Obtain 100 mL of distilled water. Fill the first test tube about two-thirds full with standard 0.01 M solution of hydrochloric acid, HCl, from the reagent shelf. Since HCl is essentially 100% dissociated into ions in dilute solution, the H^+ ion concentration of the solution is 1×10^{-2} mole/L and the pH is 2. Prepare the other solutions of pH 3, pH 4, pH 5, and pH 6 by progressive tenfold dilutions starting with the solution pH 2. To prepare the pH 3 solution use a graduated cylinder to measure 1 mL of the pH 2 solution and add 9 mL of the distilled water for a total volume of 10 mL. *Stir carefully* with a rinsed stirring rod. Remember pH 2 means 10^{-2} moles per liter H^+ ion concentration and pH 3 is 10^{-3} moles/L, so any tenfold dilution changes the pH by one unit (one exponent).

After filling the second test tube about two-thirds full, take 1 mL of the pH 3 solution and add 9 mL of distilled water to make the pH 4 solution. Take 1 mL of the pH 4 solution and add 9 mL of the water to make the pH 5 solution. Similarly prepare the pH 6 solution by diluting the pH 5 solution tenfold.

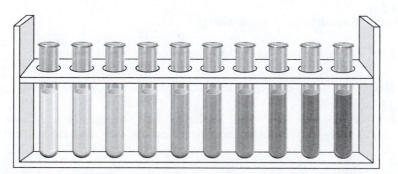

▲ **FIGURE 12.1** A set of reference indicator solutions. Note the color gradation from one end to the other.

B. Preparation of Neutral pH 7 Water

The distilled water should be pH 7. Fill the pH 7 test tube with this.[3]

C. Preparation of Solutions in the Basic Range, pH 8 to pH 12

For these start with a high pH solution, 0.01 M sodium hydroxide, NaOH, which is pH 12, and start to work down toward neutral. Dilute 1 mL of the NaOH solution with 9 mL of the distilled water to make the pH 11 solution. *Stir.* Similarly prepare solutions of pH 10, pH 9, and pH 8 by progressive tenfold dilutions, just as you did in part 1A.

Obtain your assigned indicator (in solution form) from your instructor and add 2 drops of it to each of your prepared solutions and stir. If the color is not distinct, continue to add indicator 2 drops at a time to all test tubes. Stir. Label your set, showing the indicator and the number of drops of indicator per test tube. Set it in the designated place in the room (See figure 12.1). Test tubes need not be stoppered unless they are to be left for use days later.

Note the color at each pH in all sets of indicators and record them in part 1 on the observations and results sheet. Notice that there is a gradation of color from each end to the transition zone where the color changes rather abruptly. You should be able to locate the point of greatest color change, called the transition point, to the nearest pH unit. The range of greatest color change, from just before to just after this point, is called the pH interval in table A.4.

2. Determination of the pH of Some Common Substances

Now you are ready to use the prepared indicators to determine the pH (acidity or basicity) of some common substances. Solutions of the following are suggested (try to get samples as colorless as possible so the indicator color can be seen readily).

   a.   a colorless carbonated beverage (diluted with water to half its concentration).

   b.   white vinegar (diluted tenfold).

   c.   grapefruit juice.

   d.   a colorless shampoo (diluted tenfold).

   e.   a colorless liquid laundry detergent (diluted twenty times).

   f.   household ammonia, NH_3.

   g.   baking soda (a 5% solution).

   h.   an aspirin tablet (dissolved in 20 mL of water).

   i.   a buffered tablet of aspirin (such as Bufferin dissolved in 20 mL of water).

You will need about 35 mL of each of these solutions so that you can use all five indicators one at a time with 7 mL of each substance.

To determine pH values, first pour 7 mL of the solution into each of five clean test tubes. Add 2 drops (more if needed) of each of the indicators, one indicator to each test tube, and stir. Note and record

---

[3]The distilled water is boiled to expel any dissolved CO_2 gas, which would make the water slightly acidic.

the color of each solution in part 2 of the Observations and Results sheet, and compare it with the corresponding set of reference indicator solutions, paying particular attention to the color at the transition point. You should be able to see, in each case, whether the pH of your solution is *above, below, or at the transition pH*. By combining this information for all five indicators, each of which has a different transition point, you should be able to determine the approximate pH of each substance tested. In most cases, one of the indicators will prove to be more useful than the others (the pH will be at or near the transition point). As an example, the colors for the first substance (carbonated beverage) are already indicated on the answer sheet. Note that the colors listed for the various indicators all agree with pH 5 in the table you made for part 1. (Answer Question 2, a–d on the Observations and Results sheet.)

3. Use of the pH Meter

If time permits, you should verify, using the pH meter, the pH of some of the above substances you have tested with indicators. Your instructor will calibrate the pH meter for the class by placing the electrodes in a solution of known pH and setting the dial to give that reading. Then it will be ready to test any of the solutions.

Rinse the electrodes with distilled water, place them in a beaker containing the solution to be tested, and turn on the pH meter switch. As soon as the pointer or digital reading stabilizes, read and record the pH value in part 3 of the observations and results sheet. Do this for several of the samples and compare the pH values with your previous "color" values.

Dispose of waste in the appropriate container.

QUESTIONS AND PROBLEMS

1. If a farmer is told his soil is too sour, what does this indicate about its pH? To "sweeten" the soil, should he add lime (which dissolves to make calcium hydroxide, $Ca(OH)_2$) or sulfur (which eventually makes sulfurous acid, H_2SO_3)?

2. Which would you expect to have the lower pH, lemons or oranges?

3. If a solution of hydrochloric acid of pH 2 were diluted a thousandfold, what would be the pH of the resulting solution?

4. What is the pH of a solution that is colorless in phenolphthalein and red in phenol red?

5. A solution is made by dissolving 0.06 mole of NaOH in 60 L of water. What is its pH?

6. Can an aqueous solution ever have an H^+ ion concentration of 0? What pH value would correspond to an H^+ ion concentration of 0?

EXPERIMENT

The pH Scale and
the Use of Indicators

12

OBSERVATIONS AND RESULTS

1. Color Characteristics of Some Indicators (a Class Activity)

Record your observations of the colors of the reference solutions. You can use initials or abbreviations (see next page for abbreviations).

Indicator	2	3	4	5	6	7	8	9	10	11	12	Transition Point pH
Methyl orange	___	___	___	___	___	___	___	___	___	___	___	_____
Methyl red	___	___	___	___	___	___	___	___	___	___	___	_____
Bromthymol blue	___	___	___	___	___	___	___	___	___	___	___	_____
Phenolphthalein	___	___	___	___	___	___	___	___	___	___	___	_____
Alizarin yellow	___	___	___	___	___	___	___	___	___	___	___	_____

2. Determination of the pH of Some Common Substances

Common substance	Color of Solutions for Each Indicator*					pH of Material	Acidic, Basic, or Neutral	Indicator That Was Most Useful (Nearest Transition Point)*
	MO	Phth	BTB	AY	MR			
Carbonated beverage	Y	C	Y	Y	O	5	acidic	MR
Vinegar	___	___	___	___	___	___	___	___
Grapefruit juice	___	___	___	___	___	___	___	___
A colorless shampoo	___	___	___	___	___	___	___	___

Liquid laundry
detergent
_____ _____ _____ _____ _____ _____ _____ _____

Ammonia
_____ _____ _____ _____ _____ _____ _____ _____

Baking soda
_____ _____ _____ _____ _____ _____ _____ _____

Aspirin
_____ _____ _____ _____ _____ _____ _____ _____

Buffered aspirin
_____ _____ _____ _____ _____ _____ _____ _____

*Y = yellow; C = colorless; O = orange; R = red; B = blue.

a. Is grapefruit juice as acidic as vinegar? Do your taste buds agree?

b. Calculate the hydrogen ion concentration, $[H^+]$, and hydroxide ion concentration, $[OH^-]$, in grapefruit. Show all calculations.

c. Was there an observable pH difference between buffered and regular aspirin?

d. Rank the common substances you tested from lowest to highest pH (from most acidic to most basic).

1. _____   4. _____   7. _____   10. _____

2. _____   5. _____   8. _____

3. _____   6. _____   9. _____

3. Use of a pH Meter

Record the reading of the pH meter for several of the common substances.

Common Substance	pH as Determined by pH Meter	pH as Determined by Indicators from Previous Page
_____	_____	_____
_____	_____	_____
_____	_____	_____
_____	_____	_____

The Control
of Acidity:
Buffer Solutions | 13

OBJECTIVES

1. To appreciate the importance of pH control, particularly in physiological systems.
2. To learn about the problem of maintaining close pH control in various applications.
3. To become familiar with some specific buffer systems and how they function.

DISCUSSION

Most physiological processes are extremely sensitive to pH changes. For example, the pH of human blood is maintained essentially constant at 7.2. Only at this precise pH can the blood transport oxygen and carbon dioxide adequately. If the pH drops below 7.2 (higher H^+ ion concentration), the hemoglobin in the blood will not react with oxygen, and if the pH increases (higher OH^- ion concentration), bicarbonate will not be converted to carbon dioxide in the lungs.

Fortunately, we, and our bodies, can control and maintain pH by using special types of mixtures called *buffers*. In this experiment you will prepare some buffer solutions and then observe how they work to maintain a constant pH. You will add small amounts of acid and base to buffered and un-buffered solutions and observe the results.

Weak Acids, Weak Bases, and Their Salts

A buffer system is a solution of a weak acid (or a weak base) together with one of its salts. The definition requires that we explain what is meant by "a weak acid" (or base) and "one of its salts." A weak acid or base is one that is only slightly ionized in aqueous solution. Acetic acid, $HC_2H_3O_2$, is a typical weak acid, as represented by the equation

$$HC_2H_3O_2 + H_2O \rightleftharpoons H_3O^+ + C_2H_3O_2^-$$

Mostly these
molecules exist
in solution

A few percent
of these ions
are present

Ammonium hydroxide (ammonia solution)[1] is a common example of a weak base, since it also exists to the extent of only a few percent as NH_4^+ ions and OH^- ions. Acids and bases may be classified as strong or weak, depending upon their degree of ionization. Some acids that are highly dissociated (approaching 100%) in dilute water solution are H_2SO_4, HCl, and HNO_3. Ionic bases, such as NaOH, KOH, and $Ca(OH)_2$, exist as ions in the solid state and are also completely dissociated in water solutions. On the other hand, a large number of common acids (e.g., $HC_2H_3O_2$, HCN, H_2CO_3, and H_3PO_4), organic acids (RCOOH),[2] and some bases (amines, $R–NH_2$) are only slightly ionized (a few percent or less) in water solution.

A salt of a weak acid is one that has an anion in common with the acid.[3] Such a salt can be made by allowing the weak acid to react with a base. For example, any salt which contains the acetate ion, $C_2H_3O_2^-$, is a salt of acetic acid, $HC_2H_3O_2$. A specific salt, like sodium acetate, $NaC_2H_3O_2$, can be made from the corresponding acid and base.

$$HC_2H_3O_2 \; + \; NaOH \; \rightarrow \; NaC_2H_3O_2 \; + \; H_2O$$

Similarly, sodium cyanide, NaCN, and calcium cyanide, $Ca(CN)_2$, are salts of hydrogen cyanide, HCN. Potassium monohydrogen phosphate, K_2HPO_4, is a salt of the acid potassium dihydrogen phosphate, KH_2PO_4, as indicated by the equation

$$KH_2PO_4 \; + \; KOH \; \rightarrow \; K_2HPO_4 \; + \; H_2O.$$

The salt of a weak base has a cation in common with the base. Salts of ammonium hydroxide, NH_4OH (ammonia, NH_3 solution), for example, are ammonium chloride, NH_4Cl, and ammonium sulfate, $(NH_4)_2SO_4$. Thus, a buffer solution is prepared by making a solution containing either a weak acid or a weak base plus a corresponding salt.

How Buffers Control pH

How does a buffer system function to keep the pH constant, that is, to control the H^+ and the OH^- ion concentrations? We will use a mixture of acetic acid, $HC_2H_3O_2$, and sodium acetate, $NaC_2H_3O_2$, in solution to illustrate how a buffer system works. If H^+ ion (HCl, for example) is added to the buffer system, most of the H^+ ions (hydronium ions, H_3O^+) react with $C_2H_3O_2^-$ to give weakly ionized $HC_2H_3O_2$, and so the H^+ ion concentration increases only slightly.[4]

---

[1]A solution of ammonia in water is called either ammonium hydroxide, NH_4OH, or ammonia, NH_3, solution. The overall equilibrium in the reaction of ammonia with water is

$$NH_3 \; + \; H_2O \; \rightleftharpoons \; NH_4^+ \; + \; OH^-.$$

Certainly all four of these species are present in solution. When the intended use is for the NH_3 molecules, it is most advantageous to think of it as ammonia solution. Most of the time the solution is used as a base. This probably explains why so many reagent bottles show the label as NH_4OH; hence, the designation ammonium hydroxide in this experiment.

[2]R– is used to designate any of a number of carbon-hydrogen groups.

[3]A salt of a weak acid can also be described as one whose anion is its conjugate base (i.e., the acid minus a proton). Thus, any salt containing acetate ion is a salt of acetic acid, for $C_2H_3O_2^-$ is the conjugate base of $HC_2H_3O_2$. Likewise, the conjugate acid of ammonia, NH_3, is its ammonium, NH_4^+, salt.

[4]More specifically,

$$H_3O^+ \; + \; C_2H_3O_2^- \; \rightleftharpoons \; HC_2H_3O_2 \; + \; H_2O.$$

$$H^+ + C_2H_3O_2^- \rightleftharpoons HC_2H_3O_2$$

Salt of   Weak acid
weak acid

When OH^- ion (NaOH, for example) is added to the buffer mixture the following reaction occurs.

$$OH^- + HC_2H_3O_2 \rightleftharpoons C_2H_3O_2^- + H_2O$$

The hydroxide ion concentration does not increase significantly, because almost all of the extra OH^- ions from the NaOH are removed in the reaction with acetic acid to give acetate ions and water. A buffer system, then, always contains a pair of species which are capable of capturing any extraneous H^+ or OH^- ions. In the blood, important buffer pairs are $H_2PO_4^-$ and HPO_4^{2-}, and H_2CO_3 and HCO_3^-. The following equations show how excess H^+ or OH^- ions are removed (controlled) with these buffers.

$$H^+ + HPO_4^{2-} \rightarrow H_2PO_4^-$$

Salt of   Weak Acid
weak acid

$$OH^- + H_2PO_4^- \rightarrow HPO_4^{2-} + H_2O$$

and

$$H^+ + HCO_3^- \rightarrow H_2CO_3$$

$$OH^- + H_2CO_3 \rightarrow HCO_3^- + H_2O$$

In working with pH and buffers, recall that a change of 1 pH unit means a tenfold change in H^+ or OH^- ion concentration. A solution of pH 5 has only 1% (1/100) the H^+ ion concentration of a solution of pH 3, but it has a thousand times greater H^+ ion concentration than a solution of pH 8.

In this experiment you will be measuring, fairly accurately, the pH of several buffered and unbuffered solutions by the use of indicator solutions, much as you did in experiment 12. Any changes that occur in the pH of these solutions by the addition of acid or base will be indicated by an accompanying change in color of an appropriate indicator.

Pre-Laboratory Questions | 13

1. What is a buffer solution? Why are buffer solutions important?

2. Define the terms *weak acid* and *weak base.*

3. Describe, with equations, how a solution of sodium cyanide, NaCN, with hydrogen cyanide, HCN, functions as a buffer.

4. Name some buffer pairs that are physiologically significant.

EXPERIMENT

Solutions in three pH ranges with three different indicators will be used in today's experiment. Refer to the previous experiment for a discussion of pH and indicators.

	Range	Indicator
A	(pH 3.8–5.4)	Bromcresol green
B	(pH 6.0–7.6)	Bromthymol blue
C	(pH 8.3–10.0)	Phenolphthalein

Your instructor will provide three color standards (pH 4, pH 5, and pH 6) that contain the indicator for range A, three color standards (pH 6, pH 7, and pH 8) for range B, and three color standards (pH 8, pH 9, and pH 10) for range C. Throughout this experiment you will be preparing and mixing solutions of various pH values. Keep the following in mind: always use *distilled* or *deionized* water; *mix* all solutions thoroughly; rinse all containers and test tubes with distilled water before using them; and measure volumes carefully. The experiment is designed to Illustrate how easily pH can change in unbuffered solutions compared to buffered solutions.

1. pH Changes in Unbuffered Solutions

A. Preparation of Unbuffered Solutions

Prepare, or obtain from the reagent shelf, the following solutions.

Solution 1: pure water. In a 400-mL beaker, boil about 250 mL of distilled or deionized water to expel dissolved CO_2. (Dissolved CO_2 makes a dilute solution of carbonic acid, H_2CO_3.) Allow the water to cool. Use most of it for making solution 3.

Solution 2: 0.0001 M (10^{-4} M) HCl. Mix 5 mL of 0.1 *M* hydrochloric acid, HCl, into 45 mL of distilled water (the water need not be boiled) to make 50 mL of 0.01 *M* HCl. Take 5 mL of the 0.01 *M* HCl and mix it with another 45 mL of water to make 50 mL of 0.001 *M* HCl. *Stir.* Finally, take 5 mL of the 0.001 *M* HCl and mix it with another 45 mL of water to make 50 mL of 0.0001 *M* HCl. *Stir.* This is called progressive dilution—a preferred technique for preparing very dilute solutions.

Solution 3: 0.0001 M (10^{-4} M) NaOH. Using boiled water, dilute 5 mL of 0.1 *M* sodium hydroxide, NaOH, with 45 mL of boiled distilled water to make 50 mL of 0.01 *M* NaOH. Now repeat the dilution process two more times, exactly as you did for HCl in solution 2, to obtain a 0.0001 *M* NaOH solution.

B. Determination of the pH of Unbuffered Solutions

Measure the pH of solution 1, using the indicator for range B. Fill a test tube about two-thirds full of solution 1 and add 2 drops of the indicator solution. Compare the color with the color standards and, from this, arrive at your best estimate of the pH. Repeat with solution 2, using range A indicator, and with solution 3, using range C indicator. Keep these samples for comparisons in parts C and D. Stay organized as it is easy to get test tubes mixed up. (Record results in 1A on the Observations and Results sheet.)

C. Determination of the pH of Unbuffered Solutions after the Addition of an Acid

Partially fill three more test tubes, each with one of solutions 1, 2, and 3. Add 1 drop of 1 M HCl to each and *stir*. Using the same indicators as in part 1B, determine the pH of each solution by comparison with the standards. Note any change in pH resulting from the addition of a single drop of acid. (Record results in 1B.)

D. Determination of the pH of Unbuffered Solutions after the Addition of a Base

Partially fill three more test tubes, each with one of solutions 1, 2, and 3. Add 1 drop of 1 M NaOH to each. *Stir.* Use the same indicators as before and determine the pH just as you did in part 1B. Note any change in pH resulting from the addition of a drop of base. (Record results in 1C.)

Dispose of all waste in the appropriate containers.

2. pH Changes in Buffered Solutions

A. Preparation of Buffered Solutions

Prepare, or obtain from the reagent shelf, the following solutions:

 Solution 4: $H_2PO_4^-$ and HPO_4^{2-} buffer. Mix 10 mL of 0.5 M sodium dihydrogen phosphate, NaH_2PO_4, solution with 10 mL of 0.5 M sodium monohydrogen phosphate, Na_2HPO_4, solution in a beaker or flask.

 Solution 5: $HC_2H_3O_2$ and $C_2H_3O_2^-$ buffer. Mix 10 mL of 1 M acetic acid, $HC_2H_3O_2$, solution with 10 mL of 1 M sodium acetate, $NaC_2H_3O_2$, solution.

 Solution 6: a 1:1 mixture of NH_4OH and NH_4^+ buffer. Mix 10 mL of 1 M ammonium hydroxide, NH_4OH (ammonia solution) with 10 mL of 1 M ammonium chloride, NH_4Cl, solution.

 Solution 7: a 1:4 mixture of NH_4OH and NH_4^+ buffer. Mix 5 mL of 1 M NH_4OH solution with 20 mL of 1 M NH_4Cl solution.

B. Determination of the pH of Buffered Solutions

Using the same procedure as you used for the unbuffered solutions of part 1B, determine the pH of solutions 4, 5, 6, and 7. Keep the samples identified for comparisons in the next parts. Use the following indicators:

Solution	Indicator
4	Range B
5	Range A
6	Range C
7	Range C

(Record the results in 2A.)

C. Determination of the pH of Buffered Solutions after the Addition of an Acid

Again rinse four test tubes and fill each two-thirds full with one of the buffered solutions 4, 5, 6, and 7. Add to each of these solutions 2 drops of indicator and 1 drop of 1 M HCl solution. Stir. Determine the pH of each acid-treated solution. Compare these solutions directly with the solutions in part 2B to detect any color difference. (Record results in 2B.)

D.   Determination of the pH of Buffered Solutions after the Addition of a Base

Prepare one more set of test tubes containing solutions 4, 5, 6, and 7. To each buffered solution in a test tube, add the indicator and 1 drop of 1 M NaOH solution. Determine and record the pH of each of these solutions. Compare them directly with the solutions in part 2B to detect any color differences. (Record results and answer Question 2C, 1–6.)

Dispose of all waste from part 2C and D of this experiment in the appropriate containers.

Note: Some instructors may wish to use pH meters rather than indicators for the measurement of pH either in parts of or throughout this experiment.

QUESTIONS AND PROBLEMS

1.   Explain, using equations, how human blood, which contains $H_2PO_4^-$ and HPO_4^{2-} ions, is able to maintain a nearly constant pH when base enters the bloodstream.

2.   Imagine that some digestive juices containing the strong acid hydrochloric acid accidentally entered the bloodstream through a perforated ulcer. If the blood were well supplied with $H_2PO_4^-$ and HPO_4^{2-} ions, tell what would occur chemically to keep the blood at a pH near 7.2.

3.   Blood carries dissolved CO_2. Would the presence of HCO_3^- ions in blood help to maintain a constant pH? Explain.

Name _____

Date _____ Lab Section _____

EXPERIMENT

The Control | **13**
of Acidity:
Buffer Solutions

OBSERVATIONS AND RESULTS

1. pH Changes in Unbuffered Solutions

	Solution 1: *Water*	*Solution 2:* *0.0001* M *HCl*	*Solution 3:* *0.0001* M *NaOH*

A. *Determination of the pH of Unbuffered Solutions*

Color of indicator _____ _____ _____

pH as measured by indicator _____ _____ _____

B. *Determination of the pH of Unbuffered Solutions after the Addition of an Acid*

Color of indicator _____ _____ _____

pH as measured by indicator _____ _____ _____

pH shift (change) due
to the addition of acid _____ _____ _____

C. *Determination of the pH of Unbuffered Solutions after the Addition of a Base*

Color of indicator _____ _____ _____

pH as measured by indicator _____ _____ _____

pH shift (change) due
to the addition of base _____ _____ _____

2. pH Changes in Buffered Solutions

	Solution 4: $H_2PO_4^-$ *and* HPO_4^{2-}	*Solution 5:* $HC_2H_3O_2$ *and* $C_2H_3O_2^-$	*Solution 6:* *1:1* NH_4OH *and* NH_4^+	*Solution 7:* *1:4* NH_4OH *and* NH_4^+

A. *Determination of the pH of Buffered Solutions*

Color of indicator _____ _____ _____ _____

pH as measured
by indicator _____ _____ _____ _____

B. *Determination of the pH of Buffered Solutions after the Addition of an Acid*

Color of indicator   _____ _____ _____ _____

pH as measured by
indicator   _____ _____ _____ _____

pH shift (change) due
to the addition of acid   _____ _____ _____ _____

C. *Determination of the pH of Buffered Solutions after the Addition of a Base*

1. Color of indicator   _____ _____ _____ _____

2. pH as measured
    by indicator   _____ _____ _____ _____

3. pH shift (change) due
    to the addition of base _____ _____ _____ _____

4. Summarize the pH shifts (number of pH units) due to the addition of acid or base to un-
    buffered and buffered solutions by completing the table below.

	pH shift	
	Unbuffered	*Buffered*
Acid added to acidic solutions	Solution 2 _____	Solution 5 _____
Base added to acidic solutions	Solution 2 _____	Solution 5 _____
Acid added to basic solutions	Solution 3 _____	Solution 6 _____
Base added to basic solutions	Solution 3 _____	Solution 6 _____
Acid added to nearly neutral solutions	Solution 1 _____	Solution 4 _____
Base added to nearly neutral solutions	Solution 1 _____	Solution 4 _____

Which of the unbuffered solutions, 1, 2, or 3, was the most subject to pH change? Explain why.

5. What concluding statement can you make, on the basis of your observations regarding the
    fluctuations in pH for buffered and unbuffered solutions when small amounts of acid and
    base are added?

6. In the case of solutions 6 and 7, the ratios of base to salt were different. What was the effect
    of changing this ratio?

The Concentration of an Unknown Acid: A Cooperative Approach | 14

OBJECTIVES

1. To learn a common procedure for the quantitative analysis of acids and bases.
2. To gain an understanding of mole relationships in acid-base reactions.
3. To experience a cooperative laboratory effort and become aware of the advantages and drawbacks of this approach.

DISCUSSION

This experiment on some quantitative aspects of acid-base chemistry provides an opportunity for you to simulate a professional life situation—the team approach. We suggest that you work cooperatively in teams of two. Many of the science-related professions, particularly the health care professions, use a highly structured team approach involving doctors, nurses, X-ray technicians, laboratory technologists, and pharmacists.

Frequently, decisions are based on the technical work and evaluations of others. Often team members confer before reaching a conclusion. Regardless of your chosen profession, you are likely to find it important in working with assistants and supervisors to be able to evaluate properly the data of others. It is equally important that you become aware of the reliance that others may place upon the clinical, laboratory, or other kinds of data that you provide.

Team Operation

Each partner will perform a part of the experiment and obtain data that will be used by the other partner in completing the experiment. This will make each partner dependent upon the other for the success of the experiment. If either of you makes a mistake or is inaccurate or careless, you both will experience the consequences of an unsuccessful experiment. When there are decisions to be made, you should arrive at a "group" decision. (Is a group decision always best?) If experimental results are in disagreement, the two of you must resolve the difference or find an explanation.

As you proceed through this experiment, think about the kind of interdependence that characterizes team effort. Ponder the advantages and the disadvantages of the team approach. Think of those qualities that are an asset or a liability to an individual member of a team. Also think about the experiment itself and about the degree of precision and accuracy that can be anticipated from laboratory results.

Determination of the Total Concentration of Acid in a Solution

Acidity-basicity, as expressed by pH, reflects the H^+ ion concentration—an intensity factor. However, pH does not indicate the total reservoir of acid (or base) present. To determine precisely the total concentration of an acid, including any undissociated species, a quantitative method is needed that measures the total number of moles of H^+ ions in a given volume. Titration is such a procedure. Acid-base titration is a technique in which measured amounts of acids and bases are brought together to neutralize each other. (Titrations also are used with compounds other than acids and bases as in the experiment on bleaches, experiment 8.) The object of the titration is to determine when equivalent amounts of the two reactants have been added together. The end point in an acid-base titration is the point at which equal numbers of moles of H^+ and OH^- ions have been combined to neutralize each other. An indicator that changes color at that point is used to signal when this has occurred.

Moles and Equivalents

Since some acids and bases have more than one H^+ or OH^- ion per molecule, the number of moles of acid and of base that exactly react with each other are not necessarily equal. One may be twice or three times the other. At the end point of a titration (sometimes called the equivalence point) virtually all of the H^+ ions and OH^- ions will have reacted with each other in a 1:1 ratio, including any that might have been undissociated initially. Therefore, one can deduce from the formulas of acid and base how many moles of each will be required to neutralize (the ratio of moles). For example, a mole of NaOH provides one mole of OH^- ions and a mole of H_2SO_4 provides 2 moles of H^+, so two moles of NaOH will be needed to have enough OH^- ions to neutralize each mole of H_2SO_4 (a mole ratio of 2 NaOH/1 H_2SO_4).

The number of dissolved moles in a sample can be determined by multiplying M, moles of solute per liter, by V, the volume in liters:

$$\text{No. of moles} = M \times V \text{ (in liters)}$$

Where both acid and base have equal numbers of H^+ and OH^- ions in their formulas, the number of moles of H^+ and OH^- mixed together are equal at the end point and the following equation applies:

$$M_{acid} \times V_{acid} = M_{base} \times V_{base}$$

In the example we used of H_2SO_4 which required 2 NaOH for neutralization, the above equation would need to employ a factor of 2 to the base side:

$$M_{acid} \times V_{acid} = 2(M_{base} \times V_{base})$$

Volumes may be expressed in milliliters, provided milliliters are used consistently for both volumes.

Preparation of the Standard Acid Solution

Using the foregoing equations, you will be able to determine the concentration of a solution by titrating it against a solution of known concentration. However, it will be important for you to prepare the first solution (the *titrant*) very carefully so that it is accurate. You can do this by weighing out a definite number of moles or equivalents of an acid and dissolving it to make an accurately known volume of solution. For example, if 2.0 moles are dissolved to make 4.0 L of solution, the concentration is 2.0 moles/4.0 L or 0.5 mole/L = 0.50 M. If 2.83 g of sulfuric acid, H_2SO_4, whose molecular weight is 98.0, are dissolved in water and diluted to 500 mL of solution, the molarity of the solution is

$$\frac{2.83 \text{ g}}{(98.0 \text{ g/mole})(0.500 \text{ L})} = \frac{0.0289 \text{ mole}}{0.500 \text{ L}} = 0.0578 \text{ mole/L} = 0.0578 \ M.$$

A problem often arises in finding an acid or a base that can be conveniently and accurately weighed to make the first (standard) solution. For example, sulfuric acid is not readily available in 100% purity, but only as a highly concentrated solution of imprecise composition. Sodium hydroxide is unsuitable because it is not sufficiently pure for the accuracy expected in titrations, and it is also so hygroscopic (picks up water from the air) that it increases in weight while it is being weighed. An acid or base, usually a solid of high purity, that can be weighed with great precision to make a standard solution is called a *primary standard*. Oxalic acid dihydrate, $H_2C_2O_4 \cdot 2\, H_2O$, is a reliable primary standard and will be used in this experiment.

Changes in pH during the Course of the Titration

Let us examine the pH change that occurs when a base is added to an acid. This is shown graphically in figure 14.1 for the titration of a 60-mL sample of 0.1 M strong acid (pH = 1) with 0.2 M strong base. At first the pH changes very little. Then it changes faster until, at the end point of the titration, a single drop of base is sufficient to change it by several pH units. This is the signal, provided by the color change in the indicator, to stop adding base. Any indicator that changes color in the shaded area would be suitable to signal this abrupt change in pH. When weak acids or weak bases are involved, the vertical part of the graph is shorter and the selection of a suitable indicator is more critical.

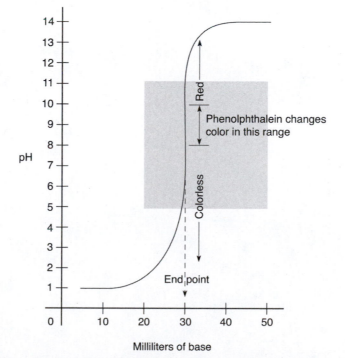

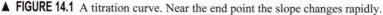

▲ **FIGURE 14.1** A titration curve. Near the end point the slope changes rapidly.

Pre-Laboratory Questions | 14

1. Very briefly describe a titration procedure.

2. What is the function of the indicator in a titration?

3. Write ionic equations for the neutralization of the following acids and bases:

   (a) $HCl + NaOH$, (b) $H_2SO_4 + NH_4OH$, and (c) $H_3PO_4 + Ca(OH)_2$.

4. How many moles of a monoprotic acid, like HNO_3, are necessary to neutralize 5 moles of NaOH? To neutralize 8 moles of $Ca(OH)_2$?

5. The molecular weight of oxalic acid, $H_2C_2O_4 \cdot 2H_2O$, is 126.0. How many grams of this acid are required to make 1 L of 0.0500 *M* oxalic acid solution (0.05 mole/L)?

6. For two professions of interest to you, describe situations in which the team approach might be used.

EXPERIMENT

This experiment is divided into three parts: (1) preparation of a standard acid, (2) preparation and standardization of a base, and (3) titration of an unknown acid. One partner will prepare the standard acid (part 1), which will be used by the other to standardize the base (part 2). Then each will complete the experiment by titrating the unknown acid to determine its concentration (part 3). You will report the results jointly. Before starting the experiment, look ahead to note the quantities that will be used in your calculations. These are the ones you will want to determine with *special care* and *precision*. In part 1, the quantities are the weight of the standard acid and the final volume of the acid solution. In part 2, the quantities are the volumes of acid and of base used, including accurate detection of the end point. In part 3, as in part 2, the measurements of the volumes are critical.

1. Preparation of a Standard Acid

You are to prepare 250 mL of a standard oxalic acid solution whose concentration (in the 0.05 M range) is to be accurately known. Review Pre-Laboratory Question 5, where you calculated the number of grams of $H_2C_2O_4 \cdot 2 H_2O$ which are needed to make 1 L of 0.5 M oxalic acid solution. To make 0.25 L of 0.05 M solution, you would need 0.0125 mole of $H_2C_2O_4 \cdot 2 H_2O$ or 1.575 g.

Weigh accurately on a piece of weighing paper approximately 1.6 g of oxalic acid dihydrate, $H_2C_2O_4 \cdot 2 H_2O$. What is meant by "weigh accurately approximately"? This means it is not important whether the weight is 1.5 g or 1.6 g or 1.7 g, but, whatever it is, the weight must be accurately known, recording as many decimal places as the balance displays. Do not waste time trying to weigh exactly 1.575 g!

Transfer the weighed acid quantitatively to a clean 250-mL volumetric flask. Add enough distilled water, with gentle swirling, to dissolve the oxalic acid. Then add just enough additional distilled water to bring the level to the scratch mark on the slim neck of the flask, making the total volume exactly 250 mL. The scratch mark has been placed so that the flask holds exactly 250 mL when filled to the meniscus at room temperature. The meniscus is the curved, shaded surface of the liquid. It is important that your eye be at the level of the meniscus and that you read just below it (see figure 1.3). Stopper the flask and mix thoroughly.

Now calculate the molarity of your acid solution. The molecular weight of oxalic acid dihydrate, $H_2C_2O_4 \cdot 2H_2O$, is 126.0. Since it has two acidic hydrogen ions per molecule, it is capable of reacting with 2 molecules (formula units) of a single base like NaOH. Find the molarity (M) by dividing the number of moles:

$$\frac{\text{grams of acid}}{126.0 \text{ g/mole}} = \text{moles}$$

by the volume of the solution, 0.250 L:

$$M = \frac{\text{grams of acid}}{(126.0 \text{ g/mole})(0.250 \text{ L})}$$

Label the flask and stopper it to prevent evaporation. This standard acid can now be used to standardize the base. (Record all calculations in part 1.)

2. Preparation and Standardization of a Base

You are to prepare about 0.5 L of approximately 0.1 *M* sodium hydroxide solution and then exactly determine its concentration by titrating it with the standard acid. With a graduated cylinder, measure approximately 25 mL of 2 *M* NaOH from the reagent shelf, and pour it into a clean bottle or flask which will hold at least 500 mL. (The graduated cylinder is not precisely calibrated and is for measuring approximate volumes. A solution labeled 2 *M* is only approximately that concentration. A precisely prepared solution would be labeled to indicate its precision, e.g., 2.000 *M*.) Dilute the 25 mL of 2 *M* NaOH with 475 mL of distilled water and mix thoroughly in the clean bottle or flask. This solution is now ready to be standardized by titration with the standard oxalic acid.

Titrations require at least one buret and a volumetric pipet. In this experiment it is suggested you add the base from the buret and measure the acid with the pipet. Both the pipet and the buret must be clean. When you have cleaned them there will be a film of distilled water on the inside that could dilute your solutions. To prevent this, you should rinse the buret with a few milliliters of the base that you plan to put into it. Likewise, rinse the pipet with the standard acid solution.

The titration will be conducted by adding an amount of base, measured with the buret, into a known quantity of acid in a 250-mL Erlenmeyer flask. Using your clean 25-mL pipet, with a squeeze bulb (figure 14.2) draw a pipet full of standard acid from your stock solution and transfer it to the Erlenmeyer flask (figure 14.3). It does not matter if the inside walls of the Erlenmeyer flask are wet with distilled water. Why?

Now that the flask holds precisely 25 mL of standard acid, your task is to find exactly how much of your base is required to neutralize it. Fill the buret nearly full with standard base solution. Drain and discard a few milliliters of the base until there are no air bubbles in the buret tip. Carefully read the buret volume at the bottom of the meniscus to two decimal places (see figure 14.4). This is your initial reading. It need not be at zero.

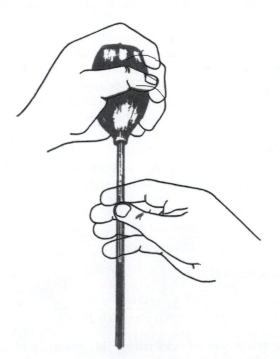

▲ **FIGURE 14.2** The proper way to fill a pipet. Apply suction with a squeeze bulb, *not* with your mouth.

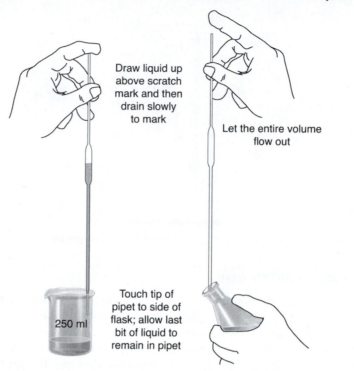

Draw liquid up above scratch mark and then drain slowly to mark

Let the entire volume flow out

250 ml

Touch tip of pipet to side of flask; allow last bit of liquid to remain in pipet

▲ **FIGURE 14.3** Using the pipet.

During the titration, when the last drop of base required for neutralization is added to the Erlenmeyer flask, the pH will suddenly increase several pH units. You need an indicator to tell you when this happens. For this purpose add 2 drops of phenolphthalein indicator to the Erlenmeyer flask. It will be colorless in acid but will turn pink the moment you have the slightest excess of base—say half a drop.

Place a piece of white paper under the flask so that the anticipated pink color can be seen more easily. While swirling the flask, add base a little at a time. Notice the localized spot of pink color where the base first hits. As long as there is an excess of acid, swirling will cause the pink color to disappear. When the pink spot begins to take more time to disappear, slow the addition; the end point is near. Add base only a drop at a time near the end point. When the last drop, or fraction of a drop, causes a pink color that persists for as long as 15 seconds,[1] the end point has been reached and the titration is complete. Read the buret and record the final volume. The volume of base used is the difference between the final and initial buret readings (figure 14.4). Calculate the molarity of the base according to the equation

$$V_a \times M_a = 2(V_b \times M_b).$$

Repeat the above procedure, using a second 25-mL sample of acid, and average the two results. If they are not close, take a third sample and average the two results in closest agreement. You may wish to confer with your partner at this point. (Record all calculations in part 2 of the Observations and Results sheet.)

Dispose of the dilute solutions by pouring down the drain with running water.

---

[1]If you swirl long enough and vigorously enough, it is possible to make the pink color fade after the end point has been reached. This is because carbon dioxide in the air, being weakly acidic, neutralizes the base.

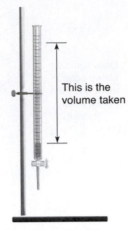

This is the volume taken

▲ **FIGURE 14.4** The end of the titration. The volume taken is the difference between the final and the original buret readings.

3. Titration of an Unknown Acid

Each partner should determine the concentration of the unknown acid, using the NaOH solution, which was just standardized in part 2. Obtain a sample of the acid solution of unknown concentration and pipet 15 mL of it into a clean Erlenmeyer flask for the titration sample. Add 2 drops of phenolphthalein indicator and titrate the unknown acid with standard base, using exactly the same procedure that you used to standardize the base. Assume the unknown acid is monoprotic (has one H^+ per molecule) and use this equation in its simplest form

$$V_a \times M_a = V_b \times M_b$$

to find the apparent molarity of the unknown acid. This time it is the concentration of base which is known and the acid which is not known. Compare the results with your partner's. If they do not agree fairly closely, each of you should run another titration. Report both your own and your partner's results, identifying your partner in the report. (Record all calculations in part 3 and answer Question 3, a–b.)

QUESTIONS AND PROBLEMS

1.   What qualities would you look for in individuals if you were selecting personnel for a clinical-scientific team upon which you would depend?

2.   If a student had a 1.5% error in the standardization of the acid and everything else was done correctly from that point on, what would be the percentage error in the final result?

3.   If the partner of the student in Question 2 made an additional 2.0% error on the standardization of the base and made no additional errors from then on, what would be the maximum and the minimum error in the final result?

4.   If there was a 0.0002-g error on each weighing and a 0.1-mL error on each volume measurement, what would be the maximum percentage error in the standardization of the base? Assume that the sample of $H_2C_2O_4 \cdot 2\,H_2O$ was 1.6 g and all volumes were 25 mL.

Name _____

Date _____ Lab Section _____

The Concentration of an Unknown Acid: A Cooperative Approach

OBSERVATIONS AND RESULTS

1. Preparation of a Standard Acid

Weight of $H_2C_2O_4 \cdot 2 H_2O$ measured

_____ g

Weight of 1 mole of $H_2C_2O_4 \cdot 2 H_2O$

_____ g

Number of moles of $H_2C_2O_4 \cdot 2 H_2O$ measured

_____ mole

Final volume of oxalic acid solution when fully
diluted to the calibration mark

_____ mL

or _____ L

Molarity of standard $H_2C_2O_4$ solution
= moles of acid/liter of solution

= _____ M

2. Preparation and Standardization of a Base

	Titration 1	*Titration 2*	*Titration 3*
Volume of standard acid	_____ mL	_____ mL	_____ mL
Molarity of standard acid	_____ M	_____ M	_____ M
Final reading of buret	_____ mL	_____ mL	_____ mL
Initial reading of buret	_____ mL	_____ mL	_____ mL
Volume of base used	_____ mL	_____ mL	_____ mL
Molarity of base used $V_a \times M_a = 2(V_b \times M_b)$	_____ M	_____ M	_____ M
Average molarity of standardized base	_____ M		

3. Titration of an Unknown Acid

	Your Titration	*Partner's Titration*
Volume of unknown acid	_____ mL	_____ mL
Final reading of buret	_____ mL	_____ mL
Initial reading of buret	_____ mL	_____ mL
Volume of base used	_____ mL	_____ mL
Molarity of base	_____ M	_____ M
Molarity of unknown acid (monoprotic) $V_a \times M_a = V_b \times M_b$	_____ M	_____ M
Average molarity of unknown acid	_____ M	
Actual value of unknown acid (from instructor)	_____ M	
Percentage error	_____ %	

Show calculations of percentage error (explained in experiment 5)

Name of partner _____

a. If the results are not in agreement with each other, or with the true (known) value, discuss some of the reasons for this. If they are in good agreement, list the places where special care was needed to avoid error in the final result.

b. List some of the advantages and disadvantages of the team approach.

> *Advantages*   *Disadvantages*

c. Would this cooperative experiment have functioned as smoothly with a group of four? Explain.

Acid and Alkali Content of Some Consumer Products | 15

OBJECTIVES

1. To determine the acid or alkali content of some important consumer products.
2. To develop critical skills toward scientific claims and to perform some typical tests to evaluate those claims.

DISCUSSION

Many consumer products contain acids and alkalis. Using the technique of acid-base titration learned in the previous experiment, you will be able to determine the concentration (as molarity, M) of the acid or alkali in these products.

Citric Acid in Citrus Fruits

The citrus fruits are a class of natural products that you will find interesting to examine. All contain varying amounts of the tricarboxylic acid, citric acid. Since acids are sour, you would probably conclude from the taste that lemons contain more citric acid than oranges. The sour taste is really due to the presence of the hydrogen ions, H^+, which are formed when citric acid $C_3H_4OH(COOH)_3$ ionizes in water.[1] As a weak acid, it ionizes only slightly.

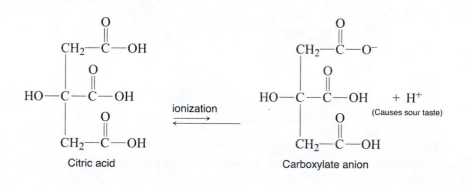

_____

[1]The hydronium ion, H_3O^+, rather than the hydrogen ion, actually is formed. The latter is used for simplicity. See experiment 12 for a discussion of these ions.

Titration of Citric Acid with Base

By titration with a standard base, the concentrations of citric acid in the juices of various citrus fruits can be determined. Sodium hydroxide reacts with all of the protons on the carboxyl groups, −COOH, of citric acid.

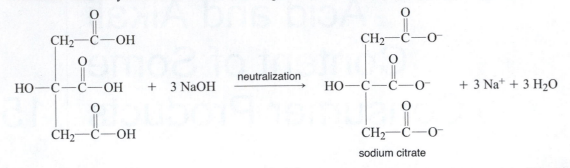

sodium citrate

Since citric acid is a triprotic acid (i.e., it has three H^+ ions per molecule), it requires 3 moles of NaOH to neutralize each mole of citric acid. You can obtain the number of grams of citric acid per liter of solution by multiplying the molarity of a citric acid solution by its molecular weight.

Antacids

Antacids are taken to neutralize excess stomach acid (about 0.1 M HCl), which is said to be a cause of upset stomach and heartburn. The reactions of some of the alkaline ingredients[2] (see table 15.1) in the commercial antacids with HCl are shown in the following equations.

$$CaCO_3 \ + \ 2\,HCl \ \rightarrow \ CaCl_2 \ + \ CO_2 \ + \ H_2O \ \ \text{(Tums, Rolaids)}$$

$$NaHCO_3 \ + \ HCl \ \rightarrow \ NaCl \ + \ CO_2 \ + \ H_2O \ \ \text{(Alka-Seltzer)}$$

$$Al(OH)_3 \ + \ 3\,HCl \ \rightarrow \ AlCl_3 \ + \ 3\,H_2O \ \ \text{(Mylanta, Maalox)}$$

$$Mg(OH)_2 \ + \ 2\,HCl \ \rightarrow \ MgCl_2 \ + \ 2\,H_2O \ \ \text{(Mylanta, Maalox, Rolaids)}$$

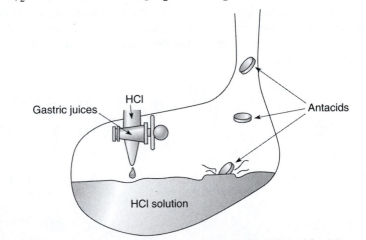

▲ **FIGURE 15.1** Hydrochloric acid in the stomach is necessary for the digestion of food; however, excess acid is injurious to the stomach.

---

[2]The bicarbonate ion, HCO_3^-, and the carbonate ion, CO_3^{2-}, are classified as bases because they react with the hydrogen ion, H^+. Unstable carbonic acid, H_2CO_3, is initially formed and then decomposes to CO_2 and H_2O.

$$\left. \begin{array}{l} HCO_3^- \ + \ H^+ \\ CO_3^{2-} \ + \ 2H^+ \end{array} \right\} \rightarrow [H_2CO_3] \rightarrow CO_2 \ + \ H_2O$$

TABLE 15.1 Alkaline Ingredients in Antacids

Alkaline Ingredient	Antacid
Aluminum hydroxide, $Al(OH)_3$	Mylanta, Maalox
Magnesium hydroxide, $Mg(OH)_2$	Mylanta, Maalox, Rolaids
Sodium bicarbonate, $NaHCO_3$	Alka-Seltzer*
Calcium carbonate, $CaCO_3$	Tums, Rolaids

*Alka-Seltzer also contains some citric acid. When it is mixed with water, the citric acid and bicarbonate react to form carbon dioxide, CO_2, accounting for the effervescence.

In today's experiment you will determine the amount of hydrochloric acid needed to neutralize a given weight of an antacid. At one time, the manufacturers of Rolaids claimed that their product neutralized 47 times its weight in stomach acid. On the basis of this experiment you may be able to verify or dispute that claim.

ENVIRONMENT, CULTURE, AND CHEMISTRY
Exact and Inexact Sciences and the Necessary Tradeoffs

Societal concerns regarding the consumer products studied in today's experiment are quite minor since the chemistry is quite straightforward, readily understood, and fairly free from controversy. Since most interfaces between quality, commercial/consumer products, and science are not so simple, our government employs monitors in an attempt to assure that quantities, purity, and advertising claims meet the established regulations. Chemistry is quite an exact science, both mathematically and intellectually, so it is natural that where chemistry intersects with phenomena studied by the less exact sciences such as weather, physiological factors, safety, environmental data, and changes in the economy, the conclusions are often blurred, and sharp differences of opinion are bound to occur. Activists easily prey on peoples' fears, often over the perceived uncertainties of science; for example, contributing to controversies over water fluoridation, vaccination of children, certain pharmaceutical drugs, and many fertilizers and pesticides. Activists are particularly dangerous when they are Hollywood celebrities, who often have a huge audience, disproportionate influence, and meager scientific knowledge. Trial lawyers purposely generate widespread concerns, both justified and unjustified, over consumer products: consider asbestos (mostly justified), pharmaceuticals (a mixed bag of opinions), automobile safety (frequently on the mark), and food product safety (regulation of probably not tough enough), since so much of our food is now coming from foreign countries, bringing with it toxic pesticides, dangerous bacteria, and damaging insects. Consider the recent destruction of Florida's orange groves by the "greening effect." The insect responsible for the destruction arrived from Asia on imported fruit. Continuing along the same line of thought, government policy makers authorize huge sums for marginal energy fixes: ethanol from corn to replace hydrocarbons (the energy realized barely covers the energy expended to make ethanol); high-speed trains (again marginal economics in addition to poor connections from one city to another); highly expensive CO_2 controls on emissions with unconvincing evidence presented for improvement in air quality or significant reductions in effects on global warming (and whenever the slightest doubt exists in the evidence, the public will tend to reject economic sacrifice). Construction of nuclear power is on hold pending political considerations on nuclear waste disposal (to the nation's detriment, politicians from both parties have said for many years, "not in my backyard"), while at the same time France successfully obtains most of its power from nuclear generation. What is at the heart of many of our festering and continuing problems: activists and politicians who refuse to acknowledge that most issues, small and large, must be dealt with as tradeoffs between the ideal and the necessary, often without the best scientific data. Educated, courageous and pragmatic leadership at all levels of government is desperately needed to sort out our nation's and the world's complex problems and to articulate the proper and most reasonable course of action to the public.

Pre-Laboratory Questions | 15

1. Name and give the structure of the acid or base in the following consumer products: (a) Tums antacid, (b) lemon juice, and (c) Rolaids antacid.

2. List the names of the indicators that will be used in the titration of each of the consumer products in this experiment. Indicate the color changes that occur at the end points.

3. Calculate the molecular weight of citric acid. Show all of your calculations.

4. A student determined that the molarity of citric acid in a citrus juice was 0.030 M. How many grams of citric acid (molecular weight = 176) are present in a 120-mL serving, assuming its density is the same as that of water?

5. Write a net ionic equation for the neutralization of sodium bicarbonate with hydrochloric acid.

EXPERIMENT

You will need some standard acid (of a known concentration) and some standard base. Accurately known concentrations of approximately 0.1 *M* are suitable.[3] Standard hydrochloric acid, HCl, solution will be provided on the reagent shelf. Standard sodium hydroxide, NaOH, solution should be available from the previous experiment, or it will be provided.

1. Citric Acid Content of Citrus Fruits

Your instructor will assign a different citrus juice (pure, unsweetened) to each of three groups in your laboratory; the suggested juices are lemon, orange, and grapefruit. Obtain a 30-mL sample of the juice assigned to you. Your task is to titrate a sample of the citrus juice to the phenolphthalein end point, using the standard base. Review the instructions in experiments 8 and 14 for appropriate titration procedures.

Obtain a clean 50-mL buret and rinse it twice with a few milliliters of standard NaOH solution (so the water remaining in the buret does not dilute the concentration). Fill it with the standard base and drain off enough to remove the air bubbles in the tip. You should have approximately 48 mL of base in your buret. Read the initial liquid volume below the meniscus. (Record in part 1 of the Observations and Results sheet.)

Obtain a clean 5-mL pipet and rinse it with a few milliliters of the citrus juice to be measured. Now pipet exactly 5 mL of the juice into a 250-mL Erlenmeyer flask. Add about 25 mL of distilled water. This dilution does not change the number of moles of acid but gives a more workable volume for titration. Add 4 drops of phenolphthalein indicator, place a piece of white paper under your flask, and begin adding the base with frequent swirling. When 1 additional drop produces a noticeable color change which persists for at least 30 seconds, you have added sufficient base, within a fraction of a drop, to neutralize the citric acid, and the titration is complete. (Record in part 1 of the Observations and Results sheet.)

Phenolphthalein changes from colorless in acid to pink in base. However, since the citrus juices are more or less colored, you should anticipate the following color changes at the end point.

> *Orange juice:* the color changes from yellow-orange to reddish orange.
> *Lemon juice:* the solution becomes darker green as the titration proceeds, with a change from green to reddish orange at the end point.
> *Grapefruit juice:* the color change is from slightly yellow to pink.

Read the final volume on the buret, and subtract the initial volume to get the volume of base added. Record on part 1 of the observations and results sheet and calculate the molarity of the citric acid, using $V_a \times M_a = 3(V_b \times M_b)$ (see experiment 14). Repeat the titration a second time with the same citrus juice, and average your results.

The instructor will collect the results from the members of your class for all of the citrus samples. Average the molarity values from the class for the citrus juice you used and compare your own result with that average. (Answer Question 1, a–d.)

> **Dispose of all waste in the appropriate container.**

---

[3] Accurately…approximately may seem contradictory. It means the quantity or concentration should be something near that suggested, but it must be accurately known.

2. Alkali Content of a Stomach Antacid

This portion of the experiment will differ from the previous parts in that you will be using standard acid in the buret and the base will be in solid form (the antacid). The alkali sample will be measured gravimetrically (weighed) rather than volumetrically. Obtain a tablet of Rolaids antacid and crush it in a mortar with a pestle until it is a uniform powder. Weigh accurately about 0.2 g of the powder on a piece of weighing paper (record in part 2 of the observations and results sheet). Carefully transfer the weighed sample to a 250-mL Erlenmeyer flask, and dissolve it in 25 mL of distilled water. (Some antacids do not dissolve completely in water until the acid is added, but they will still titrate satisfactorily.)

Clean the buret by rinsing it first with distilled water and then with two small portions of the standard acid. Now fill the buret with standard acid; the molarity of the acid should be very near to 0.1 M to simulate stomach acid. A good indicator for this experiment is methyl orange. Add 2 drops of it to the base in the flask and titrate with the standard HCl. At the end point the indicator will turn from yellow in base to pink in acid; the pink color must persist for 30 seconds. (Record in part 2 of the Observations and Results sheet.)

Now determine the weight of stomach acid that was neutralized by the weight of your Rolaids sample and see if the manufacturer's claim was verified. Assume that the density of the dilute hydrochloric acid used in the experiment is essentially the density of water, 1.0 g/mL. If the concentration of the acid is other than 0.1 M, you will need to multiply the acid volume by the factor $M/0.1$. Titrate the base in either Tums or Maalox antacid, using the same procedure that you used for Rolaids. (Record in part 2 of the Observations and Results sheet and answer Question 2, a–c.)

QUESTIONS AND PROBLEMS

1. Household ammonia purchased in the grocery store is usually 5% ammonia, NH_3, by weight. Assuming that its density is the same as that of water, what is the molarity of the solution?

2. Exactly 2 mL of 0.200 M base was required to neutralize 11.00 mL of a citric acid solution (triprotic). What was the molarity of the acid solution?

3. A new antacid tablet, containing sugar and mint flavor, was found to neutralize 32 mL of stomach acid (0.1 M HCl). How many tablets would be required to neutralize 1 mole of acid (1 L of 1 M)?

4. Describe how you would determine the amount of lactic acid in 1 L of sour milk.

5. Write molecular, ionic, and net ionic equations for the neutralization of magnesium carbonate, $MgCO_3$, with HCl.

6. Do you think that the ripeness of citrus fruit has an effect on the acid content? What leads you to believe so?

Name ______________________________

Date ______________ Lab Section ______________

EXPERIMENT 15

Acid and Alkali Content of Some Consumer Products

OBSERVATIONS AND RESULTS

1. Citric Acid Content of Citrus Fruits

Citrus fruit used ______________________________

	Run 1	*Run 2*
Volume of citrus juice	________ mL	________ mL
Final buret reading (base)	________ mL	________ mL
Initial buret reading (base)	________ mL	________ mL
Volume of standard base	________ mL	________ mL
Molarity of standard base	________ *M*	________ *M*
Molarity of citrus juice $V_a \times M_a = 3(V_b \times M_b)$	________ *M*	________ *M*

Average molarity of your citrus juice ________ *M*

Class average reported for your citrus juice ________ *M*

a. What order did you anticipate for the acidities of the citrus juices? What was the actual order as determined by the members of your class?

b. Calculate the molarity of your citrus juice. Show all calculations.

c. How far did your value for the molarity of your citrus juice deviate (differ) from the class average?

d. Using the molarity that you calculated, determine the grams per liter and the percentage of citric acid in your sample of citrus juice. Assume the juice has the same density as water. Show all calculations.

2. Alkali Content of a Stomach Antacid

Name of commercial antacid used	_____Rolaids_____	_____
Weight of antacid (base)	_____ g	_____ g
Final buret reading (acid)	_____ mL	_____ mL
Initial buret reading	_____ mL	_____ mL
Volume of standard acid	_____ mL	_____ mL
Molarity of standard acid	_____ M	_____ M
Milliliters of acid required to neutralize 1 g of antacid	_____ mL	_____ mL
Grams of acid solution neutralized per gram of antacid	_____ g	_____ g

a. Does your experimental result support the claim that Rolaids neutralizes 47 times its weight in stomach acid (about 0.1 M)? Explain.

b. Compare the neutralizing capabilities of Rolaids and the other antacid that you analyzed by calculating which will neutralize the most stomach acid per gram of antacid. How much more will one neutralize than the other?

c. List some considerations in addition to capacity to neutralize that might be factors in selecting a good stomach antacid.

Some Physiologically Important Ions: Identification of an Unknown Salt | 16

OBJECTIVES

1. To become familiar with some characteristic reactions of a number of cations and anions of physiological significance.
2. To acquire knowledge of and experience in using qualitative tests in the identification of an unknown compound.
3. To master some principles of qualitative analysis.

DISCUSSION

At least a score of metal and nonmetal ions are important in biological and physiological processes, though often they are present only in trace amounts. Although the chemistry of the body is primarily the chemistry of organic molecules, an assortment of inorganic ions is responsible for controlling the flow of fluid through cell walls, monitoring the pH of body fluids, enabling oxygen to be delivered to metabolic sites throughout the body, and permitting energy to be supplied to contracting muscles.

In today's experiment you first will become familiar with the identifying tests for a number of cations and anions. Then you will be given an unknown soluble salt, which will consist of one cation and one anion. By performing the tests you have learned, you will be able to identify both ions and, thus, the salt. The use of chemical tests to identify an unknown sample is called *qualitative analysis*.

Each kind of ion, positive or negative, has its own peculiar set of properties—size, shape, charge, and color. As a rule, the properties of a given ion are independent of the oppositely charged ion with which it is paired. Each ion reacts chemically in its characteristic way, and ions in solution generally move about independently of each other. In a solid salt the cations and anions also have their own individual properties, but they cannot migrate (unless the solid is melted or dissolved).

Knowledge of the properties of individual ions leads to knowledge of their salts. Some of the main properties of ions are the chemical reactions that can be used to identify them. Keep in mind that ions of interest always are accompanied by ions of the opposite charge, which may or may not be of interest. A few of the more important ions and their identifying chemical tests are considered here.

Cations

Iron

Iron, as the iron(II) ion, Fe^{2+}, is the central atom in the porphyrin ring of the huge hemoglobin molecule, the protein that carries oxygen in the blood. Iron is also present in the porphyrin rings of the cytochromes, which are involved in the oxidation-reduction reactions of respiration. Iron is easily identified by the blood-red color of the $FeSCN^{2+}$ complex ion, which is formed by adding thiocyanate ion, SCN^-, to iron(III) ion, Fe^{3+}, in acid solution.

$$Fe^{3+} \ + \ SCN^- \ \rightarrow \ \underset{\text{(Blood-red)}}{FeSCN^{2+}}$$

Sodium

Sodium chloride, NaCl, ordinary table salt, is present in extracellular fluids. The concentrations of sodium chloride solutions influence the passage of fluids through membranes and cell walls. The relative concentrations of sodium ion, Na^+, and potassium ion, K^+, across the membrane in a nerve cell are responsible for nerve impulse transmission. Sodium ion is identified by the intense yellow color it produces in a flame test.

Potassium

In addition to its involvement in nerve impulse transmission, potassium ion, K^+, affects the rate of synthesis of aldosterone, a component of the adrenocortical hormones. Aldosterone, in turn, controls sodium ion transport through the tubular cell membrane in the kidney. Potassium ion is identified by a violet color in a flame test.

Calcium

Calcium ion, Ca^{2+}, is present in bone as calcium phosphate, $Ca_3(PO_4)_2$, and is also involved in blood clotting. The absorption of calcium ion from the gastrointestinal tract is promoted by vitamin D and regulated by the parathyroid hormone. Calcium ion can be identified by precipitation as calcium oxalate, CaC_2O_4.

$$Ca^{2+} \ + \ C_2O_4^{2-} \ \rightarrow \ \underset{\text{(White)}}{CaC_2O_4(s)}$$

It also gives an orange-red color in a flame test.

Zinc

Zinc ion, Zn^{2+}, is an important cofactor in several enzymatic reactions. Enzymes are high-molecular-weight proteins that function as catalysts in complex reactions in the body. Many enzymes depend on cofactors for their catalytic activity. A cofactor may be either a metal ion or a complex organic molecule (coenzyme) that aids the enzyme in its function. For example, zinc ion is the cofactor along with the coenzyme NAD^+ for the enzyme alcohol dehydrogenase, which catalyzes the oxidation of ethyl alcohol to acetaldehyde.

$$CH_3CH_2OH \ + \ (O) \xrightarrow[NAD^+,\ Zn^{2+}]{\text{alcohol dehydrogenase}} \ CH_3\overset{\displaystyle O}{\overset{\displaystyle \|}{-C}}-H \ + \ H_2O$$

The oxidizing agent is represented by (O). Zinc ion can be identified by precipitation as zinc sulfide, ZnS, which is the only white insoluble sulfide.

$$Zn^{2+} \; + \; H_2S \; \rightarrow \; ZnS(s) \; + \; 2H^+$$
$$\text{(White)}$$

Ammonium

Ammonium ion, NH^{4+}, along with urea, is excreted as an end product in the metabolism of proteins and therefore is normally found in large concentration in the urine. By the addition of hydroxide ion, OH^-, the ammonium ion is converted to ammonia and water. Ammonia can be detected by its odor, or by the fact that it turns litmus blue.

$$NH_4^+ \; + \; OH^- \; \rightarrow \; NH_3(g) \; + \; H_2O$$

Anions

Chloride

The role of chloride ion, Cl^-, in the body has already been mentioned in the discussion of the sodium ion. Chloride ion is identified by precipitation as silver chloride according to the equation

$$Ag^+ \; + \; Cl^- \; \rightarrow \; AgCl(s)$$
$$\text{(White)}$$

Phosphate

The presence of phosphate ion, PO_4^{3-}, and calcium ion in bone has been mentioned already. Phosphate ion also reacts with adenosine diphosphate, ADP, to give adenosine triphosphate, ATP. ATP is the "high-energy molecule" that is required, among other things, to power many endothermic chemical reactions and to cause muscles to contract. Phosphate ion is identified by precipitation as the yellow compound ammonium phosphomolybdate, $(NH_4)_3PO_4 \cdot 12\,MoO_3$.

$$3\,NH_4^+ \; + \; 24\,H^+ \; + \; PO_4^{3-} \; + \; 12\,MoO_4^{2-} \; \rightarrow \; (NH_4)_3PO_4 \cdot 12\,MoO_3 \; + \; 12\,H_2O$$

Sulfate

The amino acid cysteine contains sulfur in a mercaptan group, $-SH$. When cysteine is metabolized by the cells, the sulfur in the mercaptan group is converted to sulfate ion, SO_4^{2-}, which is excreted in the urine. Sulfate ion is usually identified by precipitation with barium ion, Ba^{2+}, to give barium sulfate, $BaSO_4$.

$$Ba^{2+} \; + \; SO_4^{2-} \; \rightarrow \; BaSO_4(s)$$
$$\text{(White)}$$

Carbonate and Bicarbonate

Carbon compounds, such as carbohydrates, are oxidized in the cells to carbon dioxide, CO_2. Carbon dioxide from the cells is transported in the blood as a mixture of carbonate ion, CO_3^{2-}, bicarbonate ion, HCO_3^-, and carbonic acid, H_2CO_3, to the lungs, where it is exhaled as CO_2. Bicarbonate ion is the principal component in the mixture. In addition to being the method for the transportation of carbon dioxide, these chemicals form one of the important buffer systems in the blood. When acid is added to either carbonate or bicarbonate, the evolution of the odorless carbon dioxide gas serves as a means of identification.

$$2\,H^+ \; + \; CO_3^{2-} \; \rightarrow \; CO_2(g) \; \rightarrow \; H_2O$$

ENVIRONMENT, CULTURE, AND CHEMISTRY
Some Problem Ions

This experiment focuses on ions that are of physiological benefit to humans, but there are other soluble ions that must be prevented from entering the body because of their toxicity. Cyanide ion, CN^-, is an example. This ion does not occur naturally in the environment. It can be added as a poison, as was done several years ago to containers of the analgesic Tylenol. Several people died before the product could be recalled from the market. Cyanide combines with and disables certain metal-containing molecules such as hemoglobin, which carries oxygen in the blood. ($H^+ + CN^-$ gives HCN, hydrogen cyanide, the gas used in prison gas chambers.) Arsenic, an element in nature, is known in many forms ($HAsO_3^{-2}$, H_3AsO_3, AsO_4^{4-}, As_2O_3, As_2O_5, and others) and, like cyanide, has caused many deaths, both intentional and unintentional. Unintentional deaths and serious illnesses result mainly from contact with arsenic in one of the following ways: incorporation in medicines (an early treatment of syphilis for example), tonics, presence in paint and wallpaper, pesticides, wood preservative, and a contaminant in drinking water. The EPA and WHO recommend a limit of 10 ppb of arsenic (present primarily as $HAsO_3^{2-}$ and H_3AsO_3) in drinking water. Many countries including parts of the United States exceed this limit. Mercury, Hg, is a toxic material that occurs naturally and has been introduced into the environment by careless disposal procedures. It appears in our environment in several forms, including elemental Hg (a dense liquid) and its vapor, two ions Hg_2^{2+} and Hg^{2+}, and organic mercury ions, such as methylmercury, CH_3Hg^+; all of these are highly poisonous.

Mercury is a classic example of a cumulative poison. Whereas most toxins are gradually eliminated from the body, mercury is removed very slowly, over months or even years. Mercury affects most organs in the body, and also affects the central nervous system, the endocrine system; over long mercury exposure can result in brain damage. Mercury enters the body primarily through fish in the diet, although some is inhaled from the effluent of coal-fired generators, where mercury is present in the coal prior to burning. Amalgam fillings for teeth (the amalgam is 50% mercury) are a source of concern. Sweden has prohibited the use of this filling material. Fluoride ion, F^-, can be both a problem and extremely useful; it depends on the concentration. At high concentrations of approximately 3 mg per kg of body mass it is lethal, but at much lower concentrations (about 1 ppm), fluoride ion is useful in preventing tooth decay. Many cities around the world treat drinking water with low levels of fluoride. Although there is no question about the positive effect on teeth for both young and old, there has been considerable opposition to water treatment: some worry about the toxic effect of fluoride even at the lowest level; others do not support government involvement (control) in forced medication; still others (who lack insight into fluoride chemistry) do not distinguish between the benign F^- ion and the extremely dangerous (explosive and toxic) fluorine gas, F_2. Fluoride, F^-, is present in most toothpastes, and excessive use of toothpaste before age six, when the adult teeth have formed, can lead to fluorosis, a condition in which the teeth develop white and orange spots and can become permanently mottled. In March 2011, radioactive ions (isotopes) of cesium, $^{137}Ce^+$, strontium, $^{90}Sr^{2+}$ and iodine, $^{131}I^-$ (superscripts on the left indicate masses of the ions) were blown into the atmosphere by the explosion at the nuclear power plants in Japan. These ions, at sufficient concentrations, are extremely dangerous: radioactive cesium mimics potassium in the body and can cause cancer, blood diseases, and birth defects; radioactive strontium accumulates in bones and teeth, causing bone cancer and leukemia; radioactive iodide, I^-, can be inhaled or ingested, leading to thyroid cancer.

Pre-Laboratory Questions | 16

1. Why is it common practice in qualitative analysis to rinse test tubes with distilled water?

2. Which ions are commonly identified by flame tests?

3. Give the physiological significance of the ions Zn^{2+}, Fe^{2+}, Cl^-, NH_4^+, and PO_4^{3-}.

 **EXPERIMENT**

You will need a number of clean test tubes arranged in a test-tube rack or placed in a beaker. A selection of acids, bases, and salts in solution is available on the reagent shelf. In most cases you will select an ionic reagent because you need one of its ions. The other ion may be considered to be a noninterfering spectator ion. Testing the reactions of these ions in a systematic way consitutes the basis of your identifying methods. Use small amounts of reagents—usually only a few drops. Since in qualitative analysis you are testing for the presence of ions (the quality) rather than the amount present (the quantity), you must scrupulously avoid contaminating your sample and the reagents. **Always use clean test tubes rinsed with distilled water.** Water from the tap, though probably biologically pure, is not chemically pure. It is almost certain to contain as dissolved minerals some of the ions for which you are testing, such as Fe^{2+}, Ca^{2+}, Cl^-, and HCO_3^-. Thus, if you were to use tap water, these ions would be present whether or not your unknown contained them.

1. Identification of Cations

The following preliminary tests are to be made on solutions known to contain the cation you are testing for. Therefore, each test should result in a positive identification of the ion.

A. *Iron*

To about 1 mL of 0.1 M iron(III) sulfate, $Fe_2(SO_4)_3$, in a test tube[1] add a few drops of 0.1 M potassium thiocyanate, KSCN. Note the color. To another test tube, containing 1 mL of 0.1 M iron(II) sulfate, $FeSO_4$, add 3 drops of 6 M sulfuric acid, H_2SO_4, and a few drops of 3% hydrogen peroxide, H_2O_2. Now add drops of 0.1 M KSCN until you see the blood-red color of the $FeSCN^{2+}$ complex ion, which identifies iron. The purpose of the H_2O_2 is to oxidize any iron(II) ion to the iron(III) state making the test applicable to the detection of iron in either oxidation state.

The answers for the tests on iron are already given on the observations and results sheet, so that you can confirm that you are proceeding correctly.

> **Dispose of all waste in the appropriate container.**

B. *Sodium*

About the only satisfactory test for either sodium ion, Na^+, or potassium ion, K^+, is the flame test. To make the test, take 1 mL of 0.1 M sodium nitrate, $NaNO_3$, and acidify it with a few drops of 6 M hydrochloric acid, HCl. The HCl is added to assure the presence of a relatively volatile sodium salt, NaCl. Next, clean a Nichrome (or platinum) wire by first dipping it, while it is glowing hot, into another test tube of 6 M HCl and then putting it into the flame to vaporize the adhering solution. Do this several times until the flame is nearly colorless. Dip the clean wire into the solution to be tested for sodium and put the hot wire into the flame as before. As the solution is vaporized, the sodium ions give a bright yellow color to the flame. *A bright flame is a definite test. A faint yellow flame is not a positive test for sodium*; it may result from small amounts of sodium impurities, for example, sodium from dissolved glass. (Record data in 1.B on the Observations and Results sheet.)

---

[1] Mark the 1-mL level on each test tube with a marker so that it will not be necessary to measure with a graduated cylinder for each test.

C. *Potassium*

The procedure for the flame test for potassium ion, K^+, is the same as for sodium. Potassium ion gives a violet flame that does not persist for long. Take 1 mL of 0.2 M potassium nitrate, KNO_3, solution, acidify it with a few drops of 6 M HCl solution, and, after cleaning the wire, make the flame test. If you have a sample containing both sodium and potassium ions, it is necessary to filter out the bright color of the sodium yellow by looking through a piece of cobalt blue glass. Only the fainter violet of the potassium flame is visible through the cobalt glass. Use a cobalt glass and try the flame test on a solution containing both KNO_3 and $NaNO_3$. (Record data in 1C on the Observations and Results sheet.)

D. *Calcium*

Pour 1 mL of 0.1 M calcium nitrate, $Ca(NO_3)_2$, solution into a test tube and add a few drops of 0.1 M sodium oxalate, $Na_2C_2O_4$, solution. A white precipitate of calcium oxalate, CaC_2O_4, indicates the presence of calcium.[2] To another test tube containing 1 mL of 0.1 M $Ca(NO_3)_2$ add a few drops of 6 M HCl and perform a flame test on this solution. Watch for an orange-red flame. (Record data in 1D.)

E. *Zinc*

Add several drops of 6 M ammonium chloride, NH_4Cl, solution and 1 drop of 6 M ammonium hydroxide, NH_4OH, to 1 mL of 0.1 M zinc nitrate, $Zn(NO_3)_2$, solution in a test tube. The NH_4Cl and NH_4OH are a buffer pair that assures the correct pH for this test. Add a few drops of 1 M thioacetamide, CH_3CSNH_2, and heat gently for a minute or two. The thioacetamide reacts with the water (hydrolyzes) to form hydrogen sulfide, H_2S. This, in turn, reacts with Zn^{2+} to form the white precipitate zinc sulfide, ZnS, which identifies zinc. (Record data in 1E.)

F. *Ammonium*

To 2 mL of 0.1 M ammonium nitrate, NH_4NO_3, solution in a test tube add a few drops of 6 M sodium hydroxide, NaOH, solution. Moisten a piece of red litmus paper and hold it in the mouth of the test tube[3] (figure 16.1). Now gently heat the solution while swirling or agitating it. Do not let it boil. Be careful to avoid pointing the test tube toward anyone. By agitating the solution you will prevent the solution from bumping and splashing out. Cautiously smell some the vapors coming out of the test tube to detect the odor of ammonia. The proper way to do this is to waft the vapors toward your nose with a wave of your hand. Do not hold the test tube under your nose.

The NH_4^+ and OH^- ions react to give ammonia, NH_3, and water. The gaseous ammonia, driven off by heating, reacts with the water in the litmus paper to produce OH^- ions, which turn the litmus blue. A uniform blue color on red litmus paper is positive identification of NH_4^+ in the test solution. Spots of blue color on red litmus paper would be suspect, as too vigorous heating can cause NaOH solution to spatter onto the litmus paper, turning it blue. (Record data in 1F.)

---

[2]Throughout this experiment, a small amount of suspended material or a slightly turbid solution *does not* constitute a precipitate.

[3]The litmus paper may be "attached" to the inside of the test tube with water if the test tube walls are first washed down with a fine stream of water to remove all NaOH solution.

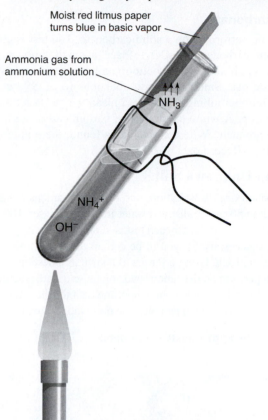

Moist red litmus paper
turns blue in basic vapor

Ammonia gas from
ammonium solution

NH_3

NH_4^+

OH^-

▲ **FIGURE 16.1** Gentle heating drives off NH_3 in the test for NH_4^+ ion.

2. Identification of Anions

These tests will be made on solutions known to contain the anion you are testing for.

A. Chloride

To 1 mL of 0.1 M sodium chloride, NaCl, solution in a test tube add a few drops of 0.1 M silver nitrate, $AgNO_3$, solution. Chloride ion combines with silver ion to form a white precipitate of silver chloride, AgCl. Verify the presence of chloride by adding a few drops (whatever is required) of 6 M NH_4OH to dissolve the precipitate, and then reprecipitate the AgCl by making the solution acidic with 6 M nitric acid, HNO_3. (Record data in 2A.)

B. Phosphate

Phosphate ion is identified by the formation of the yellow precipitate ammonium phosphomolybdate, $(NH_4)_3PO_4 \cdot 12\,MoO_3$. To 1 mL of 0.1 M sodium phosphate, Na_3PO_4, solution in a test tube add 1 mL of 6 M nitric acid, HNO_3. Add to this 1 mL of 0.5 M ammonium molybdate, $(NH_4)_2MoO_4$, solution and mix thoroughly. Place the test tube in a small beaker of boiling water for about 5 minutes. Remove the test tube from the water bath and let it stand for at least 10 minutes. The yellow precipitate, identifying phosphate, often forms quite slowly, so set the mixture aside while you continue performing other tests. (Record data in 2B.)

C. Sulfate

To 1 mL of 0.1 M sodium sulfate, Na_2SO_4, solution in a test tube add 1 mL of 6 M HCl and a few drops of 0.1 M barium chloride, $BaCl_2$, solution. A white precipitate of barium sulfate, $BaSO_4$, verifies the presence of sulfate. (Record data in 2C.)

D. Carbonate and Bicarbonate

The carbonate test should be performed on a solid carbonate. The test is characterized by the formation of a precipitate of barium carbonate, $BaCO_3$. Place a small amount of solid sodium carbonate, Na_2CO_3, in a test tube (enough to cover the bottom) and slowly add a few drops of 6 M HCl. Note the evolution of carbon dioxide gas. Smell it. Many other ions (e.g., S^{2-} and SO_3^{2-}) produce odorous gases. Using a stopper and rubber tubing (figure 16.2) lead the gas to a test tube containing 1 M barium hydroxide, $Ba(OH)_2$, solution. A white precipitate of barium carbonate formed in the second test tube confirms the presence of carbonate. Where carbonate is found, there is often bicarbonate ion, HCO_3^-. This test applies to either ion. (Record data in 2D.)

3. Identification of an Unknown Salt

Your unknown salt will contain one of the cations and one of the anions already studied in this experiment. Dissolve a gram of your salt in 50 mL of distilled water in a clean, rinsed 100-mL beaker. Use 1 or 2 mL of this test solution in place of the known ion for each test, and perform as many of the previously described tests for cations and anions as are necessary for you to be certain of the identity of your salt. The carbonate test should be performed on the solid salt. Using a format as shown in Question 3 on the observations and results sheet, list all of the tests you perform on the cations and anions, and state your conclusions, on a separate sheet of paper. Analyze these results and describe your supporting data for the salt (cation and anion) in part 3. List the unknown number, name, and formula in part 3 on the observations and results sheet.

Dispose of all waste in the appropriate containers.

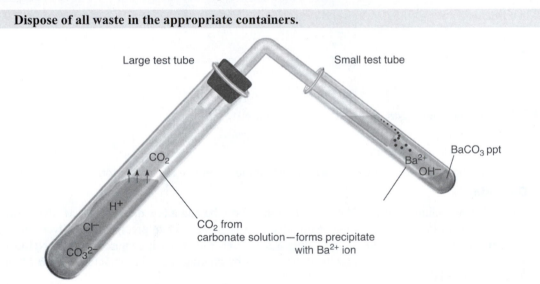

▲ **FIGURE 16.2** The test for carbonate ion, CO_3^{2-}.

QUESTIONS AND PROBLEMS

1. Nickel sulfate is green, sodium sulfate is colorless, and sodium chloride is colorless. What color would you predict for the color of nickel chloride? To which ion is the color attributable? Sodium chromate is yellow. Predict the color of nickel chromate.

2. What salt gives a precipitate with silver ion in acid solution and also produces an ammoniacal odor when treated with NaOH?

3. Which salt forms a white precipitate when heated with thioacetamide and a white precipitate when treated with barium ion in acid solution?

4. Which salt gives a yellow flame test and an odorless gas when treated with H_2SO_4?

EXPERIMENT

Some Physiologically Important Ions: Identification of an Unknown Salt

16

OBSERVATIONS AND RESULTS

1. Identification of Cations

For each cation record a brief description of the test that you made, your observations, the formula of the identifying species, and the conclusion. An example is given for iron.

Cation	Test	Observations	Formula of Identifying Species	Conclusion
A. Iron, Fe^{2+}, Fe^{3+}	Added H_2O_2 to oxide Fe^{2+} to Fe^{3+} and then added SCN^-	Solution became red in color	$FeSCN^{2+}$	Iron was present as either Fe^{2+} or Fe^{3+}
B. Sodium, Na^+				
C. Potassium, K^+				
D. Calcium, Ca^{2+}				
E. Zinc, Zn^{2+}				
F. Ammonium, NH_4^+				

Write a net ionic equation for the reaction of ammonium ion with hydroxide ion.

2. Identification of Anions

For each anion tested record information as you did for cations.

Anion	Test	Observations	Formula	Conclusion
A. Chloride, Cl^-				
B. Phosphate, PO_4^{3-}				
C. Sulfate, SO_4^{2-}				
D. Carbonate, CO_3^{2-}				

Write net ionic equations for the following. (Leave out spectator ions.)

Reaction of Ag^+ with Cl^-:

Reaction of Ba^{2+} with SO_4^{2-}:

Reaction of H^+ with CO_3^{2-}:

3. Identification of an Unknown Salt

Record the result of each test you performed on your unknown salt solution or on the solid salt. State whether the result was positive or negative. When you have made the identification, write the name and formula of your salt.

Test	Observations	Conclusion

Unknown number _____ Name and formula of salt _____

Electricity in Chemistry: Electrolytes and Electrochemistry | 17

OBJECTIVES

1. To understand the role of ionic compounds in electrical conductivity.
2. To gain competence in writing ionic equations.
3. To acquire some understanding of the relationship between electricity and matter.
4. To become acquainted with electrochemical cells and their relationship to oxidation-reduction reactions.

DISCUSSION

Long before you knew about electrons and protons, you probably were aware of a relationship between electricity and matter. For example, you may have received a shock when you touched a door handle after having walked on synthetic carpeting. Also, you have known that materials (chemicals) such as those in flashlight cells and automobile batteries (a series of cells) produce an electric current to light lamps and power motors. We shall focus upon two main electrical characteristics of matter: (1) the ability of certain types of substances to carry (conduct) an electric current and (2) the ability of certain chemical reactions to produce or use electricity.

ELECTRICAL CONDUCTIVITY

You know that metals generally are good conductors of electricity. Copper is the best of the common metals (after gold and silver). Metal atoms in the solid state are unable to move, but their outer electrons are free to migrate along the conductor (wire), and they carry the current. Another general class of compounds, called electrolytes, conducts electrical current in solution. Here the current is carried by dissolved positive and negative ions "swimming" about in the solution. Negative ions move in the same direction as do electrons in the circuit. Positive ions migrate in the opposite direction. Together the ions carry the current.

Electrolytes may be ionic solids such as potassium chloride, KCl, or they may be covalent compounds that react with the solvent to produce ions in solution. An example of the latter is hydrogen chloride gas, HCl, which reacts with water upon dissolving to give an ionized solution called hydrochloric acid.

$$HCl(g) \; + \; H_2O \; \rightarrow \; H_3O^+ \, (or \, H^+) \; + \; Cl^-$$

TABLE 17.1 Some Electrolytes and Nonelectrolytes in Water Solution

Strong Electrolytes	Weak Electrolytes	Nonelectrolytes
Sodium chloride, $NaCl$	Acetic acid, $HC_2H_3O_2$	Alcohol, C_2H_5OH
Potassium acetate, $KC_2H_3O_2$	Citric acid, $C_3H_4OH (COOH)_3$	Formaldehyde, $HCHO$
Nitric acid, HNO_3	Carbonic acid, H_2CO_3	Ethylene glycol (antifreeze),
Sodium hydroxide, $NaOH$	Ammonium hydroxide	$C_2H_4(OH)_2$
Potassium hydroxide, KOH	(ammonia), NH_4OH	Glycerine, $C_3H_5(OH)_3$
Copper sulfate, $CuSO_4$	Sulfurous acid, H_2SO_3	Acetone, CH_3COCH_3
Sulfuric acid, H_2SO_4		

Electrolytes are further classified as to whether they are strong or weak. A strong electrolyte is one that exists almost totally (approaching 100%) in the form of ions in solution. A solution of a strong electrolyte is, therefore, a good conductor. A weak electrolyte is one in which only a small fraction of the molecules in solution are dissociated into ions. Thus, a solution of a weak electrolyte is a poor conductor. In general, acids, bases, and salts are electrolytes.

We can easily test such solutions for their conductivity by placing them in an electrical circuit in such a way that a lamp glows brightly for strong electrolytes, dimly for weak electrolytes, and not at all for nonelectrolytes. In addition to being good conductors in solution, electrolytes generally have fast reactions. In fact, most ionic reactions appear to be instantaneous. Reactions of electrolytes in solution are typically written in ionic form. In such equations *strong electrolytes in solution are represented as ions, and weak electrolytes and nonelectrolytes as molecules*. In the pairs of examples shown here, notice how the net ionic equations summarize the actual chemical changes that occur in solution. The others are called spectator ions.

Ionic equation: $\underset{\text{Strong electrolyte}}{\underline{H^+ + Cl^-}} + \underset{\text{Strong electrolyte}}{\underline{Na^+ + OH^-}} \rightarrow \underset{\text{Strong electrolyte}}{\underline{Na^+ + Cl^-}} + \underset{\text{Weak electrolyte}}{\underline{H_2O}}$

Net ionic equation: $H^+ + OH^- \rightarrow H_2O$

Ionic equation: $\underset{\text{Weak electrolyte}}{\underline{HC_2H_3O_2}} + \underset{\text{Strong electrolyte}}{\underline{K^+ + OH^-}} \rightarrow \underset{\text{Strong electrolyte}}{\underline{K^+ + C_2H_3O_2^-}} + \underset{\text{Weak electrolyte}}{\underline{H_2O}}$

Net ionic equation: $HC_2H_3O_2 + OH^- \rightarrow H_2O + C_2H_3O_2^-$

Electrolytes are vital constituents of body fluids. Hydrochloric acid secreted in the stomach aids digestion. Solutions that bathe cell walls need to be in proper electrolytic balance to control the passages of fluids through the walls. Nerve impulses depend on ions to provide electrical conductivity.

Electrochemical Cells

An electrochemical cell, such as a flashlight cell, converts stored chemical energy into electrical energy. For a cell to function, it is necessary that an oxidation-reduction type of chemical reaction take place inside the cell. An *oxidation-reduction* reaction is one in which certain atoms lose electrons (*oxidation*) and other atoms gain them (*reduction*). One cannot occur without the other.

Oxidation-reduction reactions are quite common, but in order for a cell to function the reaction must occur in such a manner that the oxidation portion takes place at one location (electrode) in the cell and the reduction at another. To permit electrons lost at one electrode to be available to the other electrode, it is necessary to connect them with a wire or some other external conductor. The two electrodes are called the *cathode* (where reduction occurs) and the *anode* (where oxidation occurs). Together they constitute the cell. Inside the cell there must be a pathway along which electrical charges can move so that the electricity flows in a circuit (circle)—through the cell and back out to the conducting wire. An electrolyte in the cell serves this function. The load (lamp, motor, or other resistance) is placed in the external wire portion of the circuit. A direct wire connection, without a load, would constitute a short circuit, and the cell would run down quite fast, with the generation of much heat.

In this experiment a small Daniell cell will be constructed and the voltage it produces measured with a voltmeter. The essential features of this simple cell are illustrated in figure 17.1. The cathode consists of a piece of copper metal suspended in a solution of copper sulfate, $CuSO_4$. The anode is a metallic zinc strip dipping into a solution of zinc sulfate, $ZnSO_4$. The two solutions are separated by a porous partition, which allows ions to pass through in both directions but prevents the bulk of the solutions from mixing. A wire conductor with a load completes the circuit.

When the cell is in operation (i.e., when the circuit is closed or completed) the following reactions take place.

Cathode half-reaction (reduction):
$$Cu^{2+} + 2\,e^- \rightarrow Cu$$

Anode half-reaction (oxidation):
$$Zn \rightarrow Zn^{2+} + 2\,e^-$$

Total cell reaction:
$$Cu^{2+} + Zn \rightarrow Cu + Zn^{2+}$$

These equations tell us that Cu^{2+} ions are reduced to copper atoms (which are deposited as copper metal) in the cathode compartment. In the anode compartment, zinc metal is used up and Zn^{2+} ions are produced (oxidation). Electrons produced at the zinc anode flow through the wire to the copper cathode. Inside the cell the current is carried from the cathode to the anode by a combination of SO_4^{2-} anions migrating toward the anode and Cu^{2+} and Zn^{2+} ions migrating toward the cathode.

The Daniell cell produces about 1.1 volts (V) of direct current at the expense of zinc metal and copper ions. Of course, when either the zinc metal or the Cu^{2+} ions become depleted, the reaction stops. The cell can be recharged by forcing electrons to flow in the opposite direction, from copper to zinc, by applying a larger voltage. In this case the half-reactions and the total cell reaction are reversed, and zinc metal and Cu^{2+} ions are regenerated. The recharged cell is then ready to function again.

When a cell operates spontaneously by using up chemicals to produce electrical energy, it is called a *voltaic cell*. When a cell uses an input of electrical energy to produce a chemical change, it is called an *electrolytic cell*. Both types of cells have practical value. Often a chemical change such as the plating of silver metal onto a spoon or the production of aluminum metal is desired rather than the production of energy. A battery is a series of cells. The unit cell in an automobile battery produces about 2.0 V, so a 12-V battery consists of six of these cells connected in a series. The overall cell reaction in an automobile battery is

$$Pb + PbO_2 + 2\,H^+ + 2\,HSO_4^- \rightarrow 2\,PbSO_4 + 2\,H_2O$$

It is possible to determine whether the cell is charged by testing to see if there is appreciable sulfuric acid, H_2SO_4 (H^+ and HSO_4^- ions), in the electrolytic solution. This is sometimes done by using a float called a hydrometer to measure the density of the fluid.

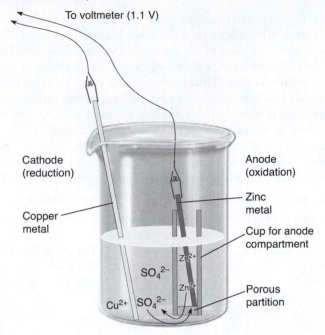

To voltmeter (1.1 V)

Cathode (reduction)

Anode (oxidation)

Zinc metal

Copper metal

Cup for anode compartment

Zn^{2+}

SO_4^{2-}

Zn^{2+}

Porous partition

Cu^{2+} SO_4^{2-}

▲ **FIGURE 17.1** The Daniell cell.

ENVIRONMENT, CULTURE, AND CHEMISTRY

Fuel Cells: Great Energy Efficiency and Less Environmental Contamination

The modern automobile with its internal combustion engine so amazingly perfected over the past century, that responds so smoothly and purrs so beautifully, is certainly an engineering marvel. However, it suffers from one great disadvantage: there is a theoretical limitation (from the Laws of Thermodynamics) on the amount of energy from the fuel that is available for propulsion; over half of it is lost irretrievably in the hot exhaust gases. This experiment is concerned with oxidation-reduction reactions, and it might be possible to use these electrochemical reactions as cells to produce electrical energy at high efficiency, much higher than is possible in the internal combustion engine. In theory any spontaneous oxidation-reduction reaction, including the burning of a fuel, can be constructed to operate as an energy-producing cell, although the actual engineering can be quite challenging. When the reaction involves a potential fuel (hydrogen, H_2, and hydrocarbons for example) with an oxidizer (such as oxygen, O_2), it is called a fuel cell. Since the fuel cell does not evolve hot gases, it does not suffer the same energy limitation as the internal combustion engine. Therein lies an enormous advantage for the energy challenges of our society. If such a fuel cell can be perfected, its potentially greater efficiency could generate electrical energy using far less of our precious, dwindling fuel resources and produce far fewer greenhouse gases, thus reducing damage to the environment. Why are fuel cells not in common use today? The answer is found in the fuel cell's many technical difficulties. A fuel cell, using the simplest fuel H_2 with O_2 to produce water, H_2O, has been made to operate quite effectively in limited situations, but there are many drawbacks for everyday usage. Hydrogen is not a primary fuel source: there are no H_2 mines or wells, so it must be made from another source such as water or methane, CH_4, and production from these sources requires significant energy. Also, H_2 is a bulky gas that cannot be liquefied at ordinary temperatures, is not easily transported, and is quite explosive. Furthermore, platinum electrodes are used in a hydrogen fuel cell, and there is concern that there may not be sufficient platinum in the world for widespread usage. The readily available complex fuels (CH_4, other hydrocarbons, and coal) have so far proven stubborn nuts to crack as relates to making workable fuel cells. (Hydrocarbon fuel cells would produce some CO_2, but far less than conventional fuels because of their greater efficiency.) Big problems exist in developing fuel cells, but thankfully there is no theoretical restriction on the maximum energy, so the goal is worth a great effort and expense. Perhaps the stakes are sufficiently high to warrant a national commitment on the scale of the Manhattan Project, which led to the development of the atomic bomb.

Pre-Laboratory Questions | 17

1. What is an electrolyte?

2. What is oxidation? What is reduction? Is it possible to have one without the other?

3. What type of chemical reaction is the basis for electrochemical cells? Name the two general types of electrochemical cells.

4. How did cations and anions get their names?

5. Write the cell reactions for (a) the Daniell cell and (b) the lead automotive battery.

EXPERIMENT

This experiment can be done by students working in pairs or by the instructor in a demonstration.

1. Conductivity of Electrolytes

An apparatus of the type shown in figure 17.2 is useful for determining the electrical conductivity of a solution. It is constructed so that the electrical circuit can be completed by the solution to be tested. The intensity of the light is a measure of the conductivity of the solution. Note the arrangement of the lamps in figure 17.2. Strong electrolytes are such good conductors of electricity that all three lamps light up. With a weak electrolyte, the current through the electrolytic solution may not be sufficient to heat all three of the lamp filaments to incandescence. In this case, unscrew the 40-watt (W) lamp so that the 10-W lamp and the 1-W glower can light up. For the weakest electrolytes, unscrew the 10-W lamp also, since the solution can conduct only enough current for the sensitive glower. Not even the glower responds to a solution of a nonelectrolyte. Very pure water conducts so poorly that it can be viewed as a nonelectrolyte.[1] Some instructors choose to use a simpler apparatus with only one lamp and note whether it glows brightly, dimly, or not at all.

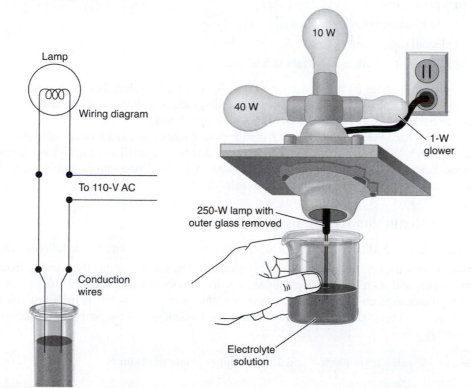

▲ **FIGURE 17.2** Apparatus for comparing the electrical conductivity of solutions, using a 110-V current source. For most solutions the 10-W lamp is appropriate. The 40-W lamp is for strong electrolytes, and the glower is for very weak electrolytes.

---

[1]Distilled water may have sufficient conductivity, due to dissolved carbon dioxide or trace impurities, to light the glower. Removing the last traces of electrolytes from distilled water is difficult. Boiling will expel most of the CO_2.

Caution: Avoid touching the electrodes when the apparatus is connected to the electrical outlet. Your fingers could become the conductor to complete the 110-V circuit. Keep your hands dry. Do not let the electrode wires touch each other. Disconnect the apparatus when you are not actually making measurements.

Test the following substances with the conductivity apparatus. Note the relative brightness of the light and classify each substance as a strong electrolyte, a weak electrolyte, or a nonelectrolyte.

   a. tap water

   b. distilled water

   c. 0.1 M sugar solution, $C_{12}H_{22}O_{11}$

   d. 0.1 M hydrochloric acid, HCl

   e. 0.1 M acetic acid, $HC_2H_3O_2$

   f. 0.1 M phosphoric acid, H_3PO_4

   g. 0.1 M sulfuric acid, H_2SO_4

   h. 0.1 M ammonium hydroxide (ammonia), NH_4OH

   i. 0.1 M sodium hydroxide, NaOH

   j. a mixture of equal volumes of *d* and *i*

   k. a mixture of equal volumes of *e* and *h*

   l. a mixture of equal volumes of *f* and *i*

   m. ethyl alcohol (anhydrous)

   n. ethyl alcohol and distilled water (about half and half)

Proceed by placing about 20 mL of the solution to be tested in a 50-mL beaker that has been rinsed with distilled water. Raise the beaker, as shown in the diagram, so that the wire electrodes dip into the contents of the beaker. Try to have the wires about the same distance apart and submerged to the same depth for all tests. Disconnect the electrodes from the power source and rinse them with distilled water before each test. Why? This can be done most easily by directing a small stream of water from a wash bottle onto the electrodes. (Record your observations in part 1 of the Observations and Results sheet and answer Question 1, a–c on the Observations and Results sheet.)

2. Oxidation-Reduction Reactions

Place about 25 mL of 0.2 M copper sulfate, $CuSO_4$, in a 50-mL beaker. Into this solution place a bright iron nail (size 8d or smaller). After a few minutes, remove the nail and examine the coating. Iron is more active than copper, and so it will replace copper from its compounds in solution. (This is discussed in experiment 6.) Place a piece of mossy zinc metal or a zinc strip in the solution for a few minutes and observe the reaction. This is the reaction that occurs in the Daniell cell of the next part. What is the reaction? (Answer Question 2, a–b.)

Dispose of all waste from parts 1 and 2 in the appropriate containers.

3. A Simple Daniell Cell

The reaction between zinc metal and copper ions can be made to take place in a cell and produce electrical energy. The two parts of the reaction (half-reactions) are physically removed from each other by constructing a cell having two compartments. This can be achieved most conveniently if one compartment is separated from the other by a porous cup or a container with a porous plate, as shown in figure 17.1. The porous partition permits ions to migrate through its walls, thus connecting the circuit, but prevents the two solutions from mixing.

To construct the cell, use a 150-mL beaker containing about 50 mL of 0.2 M copper sulfate, $CuSO_4$, solution for the outer vessel. For the inner container use a smaller porous cup, a fritted glass crucible, or a similar container filled to the same level with 0.2 M zinc sulfate, $ZnSO_4$, solution. Some containers have porous plates in the bottom only. If you use a fritted glass crucible, be certain that no air bubbles become trapped below the porous plate to cut off the flow of current. One way to eliminate the bubbles is to place an empty fritted glass crucible in the $CuSO_4$ solution and allow the solution to force the air up through the dry plate. When the porous partition becomes moist with $CuSO_4$ solution, the $ZnSO_4$ solution may be added. Insert a strip of copper metal in the $CuSO_4$ solution and a strip of zinc metal in the $ZnSO_4$ solution. The strips should be long enough to rise above the level of the solutions. Use alligator clips to attach wires to the metal electrodes, and connect the other end of each wire to a high-resistance direct current voltmeter capable of reading a few volts. Record the voltage and answer Question 3a on the observations and results sheet. Failure to get a voltage reading is probably caused by an air barrier in the porous partition. (Answer Question 3, b–j.)

Dispose of all metal solutions in appropriate containers.

QUESTIONS AND PROBLEMS

1. Arrange the following list of compounds from poorest to best conductors of electricity: copper metal, gasoline, ammonium hydroxide, hydrochloric acid, and water.

2. Which of the following reactions theoretically could be used as a basis for constructing an electrochemical cell?

   a. $Cu^{2+} + Fe \rightarrow Cu + Fe^{2+}$

   b. $Ag^+ + Cl^- \rightarrow AgCl$

   c. $Ca(OH)_2 + H^+ \rightarrow HSO_4^- \rightarrow CaSO_4 + 2\,H_2O$

   d. $I_2 + 2\,S_2O_3^{2-} \rightarrow 2I^- + S_4O_6^{2-}$

3. Write balanced ionic equations for the following reactions. Cross out the spectator ions, leaving only the net ionic equations.

   a. barium chloride + sodium sulfate $\rightarrow$

   b. $AgNO_3 + KBr \rightarrow$

   c. $NaOH + H_2SO_4 \rightarrow$

   d. $HCl + NaC_2H_3O_2 \rightarrow$

   e. sodium hydroxide + ammonium chloride $\rightarrow$

    f. $CaCO_3 + HNO_3 \rightarrow$

    g. zinc sulfide + hydrochloric acid $\rightarrow$

4.   Write the following equations in both ionic form and net ionic form.

    a. $2\,HCl + BaCO_3\,(s) \rightarrow BaCl_2\,(g) + CO_2\,(g) + H_2O$

    b. $HNO_3 + KC_2H_3O_2 \rightarrow HC_2H_3O_2 + KNO_3$

EXPERIMENT

Electricity in Chemistry: Electrolytes and Electrochemistry

17

OBSERVATIONS AND RESULTS

l. Conductivity of Electrolytes

Record your observations from the conductivity test of the various solutions and classify each substance as a strong electrolyte, weak electrolyte, or nonelectrolyte.

Solution	Brightness of Light			Electrolytic Classification
	40 W	10 W	Glower	
a. Tap water	_____	_____	_____	_____
b. Distilled water	_____	_____	_____	_____
c. 0.1 M sugar solution	_____	_____	_____	_____
d. 0.1 M hydrochloric acid	_____	_____	_____	_____
e. 0.1 M acetic acid	_____	_____	_____	_____
f. 0.1 M phosphoric acid	_____	_____	_____	_____
g. 0.1 M sulfuric acid	_____	_____	_____	_____
h. 0.1 M ammonium hydroxide	_____	_____	_____	_____
i. 0.1 M sodium hydroxide	_____	_____	_____	_____
j. A mixture of d and i	_____	_____	_____	_____
k. A mixture of e and h	_____	_____	_____	_____
l. A mixture of f and i	_____	_____	_____	_____
m. Ethyl alcohol (anhydrous)	_____	_____	_____	_____
n. Ethyl alcohol and distilled water	_____	_____	_____	_____

a. What products are formed as a result of mixing the acids and bases in *j*? in *k*? in *l*?

b. Write ionic equations and net ionic equations for the acid-base neutralizations in *j*, *k*, and *l*.

Ionic equation (j):

Net ionic equation (j):

Ionic equation (k):

Net ionic equation (k):

Ionic equation (l): $H_3PO_4 \; + \; Na^+ \; + \; OH^- \; \rightarrow \; H_2PO_4^- \; +$

Net ionic equation (l):

c. Why did mixture *k* make a brighter light than either of the starting solutions? (Look at the equation.)

2. Oxidation-Reduction Reactions

a. Describe the change in the appearance of the iron nail.

b. Tell what occurred at the surface of the nail.

c. Write an ionic equation for the reaction between iron metal and copper ions.

$$Fe \ + \ Cu^{2+} \ \rightarrow$$

d. What occurred at the surface of the zinc metal?

e. What went into solution? What was deposited from the solution?

f. Write an ionic equation for this reaction (the Daniell cell reaction).

3. A Simple Daniell Cell

a. What is the voltage of your cell as measured by the voltmeter?

b. Write equations for the half-reactions that occur at the anode and the cathode. (Remember: electrons are written in half-reactions).

Anode half-reaction:

Cathode half-reaction:

c. Which half-reaction, oxidation or reduction, occurs at the cathode? at the anode?

d. Write the total cell reaction.

e. Which electrode loses mass?

f. Which electrode gets heavier?

g. Suggest a reason why you face little or no electrical shock hazard when you work with a Daniell cell.

h. What carries the electric current through the solution?

i. Why must there be contact between the two solutions through the porous partition?

j. Make a sketch of your cell showing where reactions occur. Also indicate (with an arrow) the direction of the flow of electrons and negative ions.

Molecular Architecture: Some Comparisons Between Covalent and Ionic Compounds | 18

OBJECTIVES

1. To become familiar with the differences between covalent and ionic compounds.

2. To learn how the type of bonding and molecular structure directly affects compounds.

3. To compare the physical and chemical properties of some sets of isomers.

4. To prepare for the changes in experiment structure required when the study of organic chemistry begins.

DISCUSSION

The field of chemistry is divided into two broad categories called inorganic chemistry and organic chemistry. You may already be aware of these divisions because chemistry texts have inorganic and organic sections and chemistry departments offer both inorganic and organic chemistry courses. Since all organic compounds, but few inorganic compounds, contain carbon, we can define organic chemistry as the chemistry of carbon compounds. The definition is somewhat arbitrary but convenient and useful for organizing our knowledge of chemistry.

Most organic compounds have covalent bonds; most inorganic compounds, notably salts, are held together by ionic bonding. Although a number of covalent, inorganic (non-carbon) compounds, such as H_2O, HCl, B_2H_6, SO_2, NH_3, and PCl_3 do exist, covalence is much more typically found in carbon compounds. A few elements in addition to carbon (silicon, phosphorus, boron, and sulfur) are known to bond covalently to themselves to form short chains or rings, but carbon is so much more versatile in doing this that it is in a class by itself. This ability of carbon to bond to other carbon atoms in indefinitely long chains accounts for the almost limitless variety of organic compounds.

Most of the observed differences between typical organic and inorganic compounds are accounted for by the character of the covalent bonding of carbon, which is the emphasis of this experiment. In today's experiment you will compare some physical properties (solubilities and melting points) of covalent and ionic compounds and examine the ability of carbon compounds to form chains, rings, and isomers.

Comparisons Between Covalent and Ionic Compounds

The most pronounced physical differences between covalent and ionic compounds are generally their melting points, solubilities, and electrical conductivities. All three differences are at least partially due to the greater strength of ionic bonds. Comparisons of some properties of covalent and ionic compounds are given in table 18.1.

The strengths of the bonds between particles is responsible for the differences in melting points of covalent and ionic compounds. The van der Waals attraction that exists between molecules in covalent compounds is considerably weaker than bonds in ionic compounds (see figure 18.1). Consequently, it takes less energy (i.e., lower heat) for molecules of covalent compounds to break out of their ordered solid state and pass into a more random liquid state. In other words, covalent compounds melt at lower temperatures than salts. The same reasoning can be applied to boiling points: Almost all covalent compounds boil at lower temperatures than ionic compounds.[1]

TABLE 18.1 Comparison of the Properties of Covalent and Ionic Compounds

Covalent Compounds	Ionic Compounds
Most have low melting points (usually below 350°C). Many are liquids or gases at room temperature.	Most have high melting points (usually above 350°C, commonly to 1000°C). All are solids at room temperature.
High percentage are soluble in nonpolar solvents. Few dissolve in water or conduct electricity.	High percentage are water-soluble and conduct electricity. Few dissolve in nonpolar solvents.
Most will burn.	Essentially none burn.
Many have pronounced odors.	Few have odors.

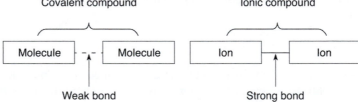

▲ **FIGURE 18.1** Relative strengths of bonds between particles in covalent and ionic compounds.

Most ionic compounds dissolve in water because the polar water molecule forms relatively strong bonds with ions. The negative "oxygen end" of the water molecule bonds with the cation (M^+) and the positive "hydrogen end" bonds with anion (X^-), as is illustrated below.

---

[1] High molecular weight polymers (plastics, proteins, starches, etc.), which contain covalent bonds, would have extremely high boiling points, but most of them fragment (decompose) into smaller molecules long before the boiling point is reached.

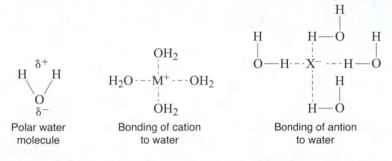

Polar water molecule

Bonding of cation to water

Bonding of antion to water

As the number of bonds between water molecules and an ion increases, the bonds between the ion and its neighboring ions in the crystal structure weaken, and eventually the free hydrated ion passes into solution. Covalent compounds are soluble in nonpolar solvents[2] but not in water unless their molecules form hydrogen bonds with water. Oxygen-containing[3] organic compounds with four or fewer carbons usually dissolve in water because of hydrogen bond formation as illustrated below.

$CH_3 - O - H$

Methyl alcohol

$CH_2 = O - - - H - O$

Formaldehyde

$CH_3 - C - CH_3$

Acetone

Chains and Rings of Carbon Atoms

The element carbon is unique in that it bonds repeatedly with itself to form stable chain and ring (also called cyclic) compounds; n-hexane, C_6H_{14}, and cyclohexane, C_6H_{12}, are examples of 6-carbon chain and ring molecules. Progressively longer chains occur in decane, $C_{10}H_{22}$, mineral oil,[4] and even n-hectane, $C_{100}H_{202}$.

$CH_3 - CH_2 - CH_2 - CH_2 - CH_2 - CH_3$

n-Hexane—a chain compound

Cyclohexane

Isomers

Isomers are molecules with the same molecular formula but different molecular structures (i.e., different arrangements of the atoms within the molecules). Isomerism is common among carbon compounds but rather unusual in other covalent and ionic compounds. Isomers normally have different physical and chemical properties, as you will observe in today's experiment. *ortho*(*o*)-Dichlorobenzene and *para*(*p*)-dichlorobenzene, $C_6H_4Cl_2$, are a pair of common isomers. n-Butyl alcohol, t-butyl alcohol, and diethyl ether are all isomers with the molecular

---

[2]Examples of nonpolar solvents are carbon tetrachloride, chloroform, hexane (and other hydrocarbons), and ether.

[3]Low molecular weight nitrogen compounds, such as amines and amides, also form hydrogen bonds and dissolve in water.

[4]Mineral oil is a mixture of hydrocarbon molecules of $C_{20}H_{42}$ and up.

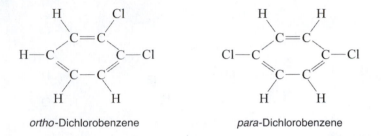

ortho-Dichlorobenzene   *para*-Dichlorobenzene

formula $C_4H_{10}O$. Since the two alcohols are members of the alcohol family (characterized by $-OH$ groups), most of their properties are similar. However, diethyl ether is a member of the ether family, which is identified by an oxygen bonded to two carbon groups, and its properties are considerably different. You will examine the properties of these three compounds in today's experiment.

ENVIRONMENT, CULTURE, AND CHEMISTRY
An Amazing Liquid

Water is an absolutely amazing molecule. Because of its very special physical properties, it has the unique ability to support life. It is far superior to any alternative. Water is a small molecule with a molecular weight of 18 amu (atomic mass units) but it boils at the unexpectedly high boiling point of 100°C. Compare this to methane, CH_4, with a molecular weight of 16 amu, very similar to water, but with a drastically lower boiling point of −161°C. This means that water will be a liquid at earth's temperature and pressure, evaporate quite slowly, cooling the outsides of bodies of animals but not forming a vapor inside their cells and bodies, which would lead to death. In addition, water dissolves a wide range of particles from the relatively small sodium and chloride ions to immensely larger proteins such as hemoglobin and a wide variety of enzymes. The high dissolving capacity of water makes it possible for the blood (mostly water) to carry a broad range of materials. Furthermore, water is also unique in that it expands on solidification, facilitating the crumbling of rocks into soil. Water owes most of these unusual properties to the simple fact that the two hydrogens and oxygen lie not in a straight line (180°), but at an angle of about 104°. Due to the bond angle, the hydrogens are somewhat positive and the oxygen has a partial negative charge. If the atoms all lay in a straight line, the charges would cancel each other and water would be a non-polar molecule; as it is, the molecule is polar, meaning that one end is positive and the other negative. This polarity means that water molecules "hang on to each other" (by means of a hydrogen bond), which raises the boiling point; it also explains why water dissolves charged particles (ions and enzymes) and $-OH$ containing molecules such as alcohols and many carbohydrates. This unique and peculiar nature of water has caused astronomers to focus on water in their search for life in the far reaches of the universe. The basic question is: can a planet be located that has a temperature (along with many other requirements) that would permit water to exist in the liquid form? If a planet with liquid water were discovered, at least simple forms of life might be found. Considering the large number of planets (about 50) that have recently been located within 80 light-years of earth, the idea of life on other planets is intriguing. If life were found on other planets, it would be expected to be based on the carbon-oxygen system. No scientist has proposed the possibility of life based on any other system, such as the closely related silicon-nitrogen system because the physical and chemical properties of these elements are so inferior for life. Also, no one has seriously suggested the development of life based on elements other than those in our periodic table. The hunt is on for a planet with liquid water.

Pre-Laboratory Questions | 18

1. Why is water called a polar molecule? Explain its dipole nature on the basis of the molecule shape.

2. List several ways in which ionic and covalent compounds differ. Which of these differences will you study in today's experiment?

3. Draw the structures for the isomers with the formula $C_3H_6Cl_2$, showing each bond as a dash. Does each isomer have the same number of bonds? What is that number?

4. Indicate which of the following compounds have ionic bonds and which have covalent bonds: $MgCl_2$, C_4H_{10}, CO_2, Li_2O, C_3H_8, PCl_3, HCl.

5. Predict which of the above compounds have high melting points and which have low melting points.

6. Draw structures of the chain (linear) butane, and the ring (cyclic) cyclobutane, showing each bond as a dash.

EXPERIMENT

1. Melting Point Comparisons

A. Covalent Compounds

Construct a melting point apparatus (see figure 18.2), and determine the melting points of naphthalene, $C_{10}H_8$, and p-dichlorobenzene, $C_6H_4Cl_2$. Some suggestions for taking the melting points are

(1)   Obtain a new capillary melting point tube, and add the powdered compound to it by pressing the open end of the tube into the sample. Then turn the tube over and tap it so that the sample falls to the bottom of the tube. You will need about 2 mm of sample in the tube. (See figure 18.3.)

(2)   Attach the tube to the thermometer with a rubber band (figure 18.3), and align the end of the tube with the mercury of the thermometer.

(3)   Heat the water bath so that the temperature rises about 10 degrees per minute. Stir the water constantly as you heat it.

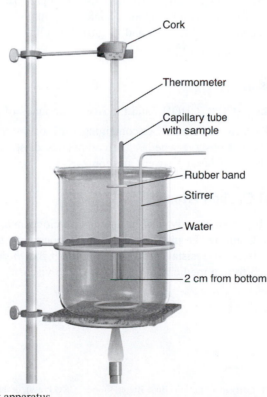

▲ FIGURE 18.2 Melting point apparatus.

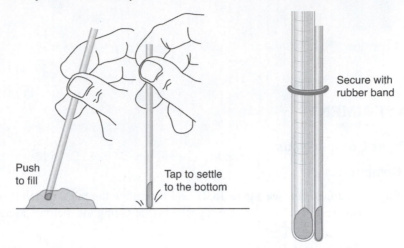

▲ **FIGURE 18.3** Filling a capillary melting point tube.

(4)   Watch the sample closely. Note the exact temperature at which it starts to melt. Record the melting point range for each compound. You may wish to retake a melting point if the temperature of the bath was rising too rapidly for you to obtain an accurate reading. Look up the reported melting points in a reference suggested by your instructor.[5] (Answer Question 1A, 1–5 on the Observations and Results sheet.)

B. Ionic Compounds

You will not be able to determine the melting points of ionic compounds with the melting point apparatus since water boils far below their melting points. Therefore, use an appropriate reference book to look up and record the melting points of the following important ionic compounds: sodium chloride, NaCl (table salt); potassium iodide, KI (additive in table salt); and magnesium sulfate, $MgSO_4$ ($MgSO_4 \cdot 7H_2O$ is Epsom salts). (Answer Question 1B, 1–3.)

2. Solubility Comparison

Determine if isopropyl alcohol, $(CH_3)_2CHOH,$ and the five compounds of part 1 are soluble in water and in methylene chloride. To each of six test tubes containing 1 mL of water, add a different compound (about the size of a pea). Stir and observe whether the compounds dissolve. Repeat using methylene chloride as the solvent. Use dry test tubes. (Answer Question 2, a–c.)

3. Chains and Rings of Carbon Atoms

Demonstration: Before you begin your study of *n*-hexane and cyclohexane, your instructor will make models of these compounds. Compare the physical properties (appearance and odor) of *n*-hexane and cyclohexane. Compare the viscosity (resistance to flow) of *n*-hexane, *n*-decane, and mineral oil by squeezing a few drops of each from a medicine dropper. Does the viscosity change with chain length? (Answer Question 3, a–f.)

---

[5]Robert C. Weast's *Handbook of Chemistry and Physics,* (Boca Raton, Florida: CRC Press), or any recent edition, is suitable. Note that the organic compounds (carbon compounds) and inorganic compounds are in separate sections of the book.

4. Isomers

Demonstration: Before you proceed to examine each set of isomers described below, your instructor will construct models to demonstrate the molecular structure. Compare the physical states (solid or liquid) of o-dichlorobenzene and p-dichlorobenzene. Note the odors of n-butyl alcohol and t-butyl alcohol by wafting the vapors. Are they identical? Determine the water solubilities of the two alcohols by adding each a drop at a time up to a limit of 15 drops each to test tubes containing 1 mL of water. Shake after each addition until no more will dissolve. That point is reached when a second layer forms and it becomes cloudy. Record the number of drops. Now compare the chemical properties of n-butyl alcohol and t-butyl alcohol. Demonstration: Your instructor will add a small piece of sodium to each alcohol in dry test tubes. Note the rate of formation of hydrogen bubbles.

Note the odor of diethyl ether and compare it with the odor of the alcohols by wafting the vapors.

Demonstration: To observe another chemical property—how well each burns (i.e., reacts with oxygen)—your instructor will ignite a few drops of each compound on a spatula. Note and record how quickly each burns away. (Answer Question 4, a–k.)

Dispose of all waste in the appropriate containers.

QUESTIONS AND PROBLEMS

1. Predict which would have the higher melting point, calcium chloride, $CaCl_2$, or acetyl chloride,

    $$CH_3-\overset{\overset{\displaystyle O}{\|}}{C}-Cl.$$ Give an explanation for your prediction.

2. The odor of liquid (melted) p-dichlorobenzene is substantially stronger than the odor of the solid. Explain.

3. Give an explanation for the insolubility that you observed for naphthalene in water.

4. Two other butyl alcohols, besides n-butyl and t-butyl, are known. Draw the structures for these isomers.

5. Diethyl ether is partially soluble in water. Explain the role of water molecules in dissolving ether.

EXPERIMENT

Molecular Architecture:
Some Comparisons Between
Covalent and Ionic Compounds

18

OBSERVATIONS AND RESULTS

1. Melting Point Comparisons

    A. *Covalent Compounds*

       1. Melting point of naphthalene   Observed   _____._____ °C

                                      Reported   _____ °C

           Melting point of *p*-dichlorobenzene   Observed   _____ °C

                                      Reported   _____ °C

       2. From the standpoint of motion of molecules, describe what happens to naphthalene at its melting point.

       3. Compare the odors of naphthalene and *p*-dichlorobenzene.

       4. Have you smelled either of them previously?

       5. What is a common use for these compounds?

    B. *Ionic Compounds*

       1. Record the melting point of the ionic compounds:

           sodium chloride (NaCl)   _____ °C

           potassium iodide (KI) _____ °C

           magnesium sulfate ($MgSO_4$) _____ °C

       2. Why are the melting points of the ionic compounds so much higher than those of the covalent compounds?

       3. Describe an important use for KI and for $MgSO_4 \cdot 7H_2O$.

2. Solubility Comparisons

a. Indicate if you found the following compounds to be soluble, partially soluble, or insoluble and write the formula for any particles present in the solution in which the compounds dissolved, or partially dissolved:

Compound	*Water*	*Methylene Chloride*	*Particle(s) Present in Solution*
sodium chloride			
potassium iodide			
magnesium sulfate			
naphthalene			
p-dichlorobenzene			
isopropyl alcohol			

b. Did any of the ionic compounds dissolve in the nonpolar solvent methylene chloride? Were any covalent compounds water-soluble? Explain affirmative answers using bonding considerations.

c. Describe the role of water in dissolving *one* of the ionic compounds.

3. Chains and Rings of Carbon Atoms

a. Do *n*-hexane and cyclohexane have identical odors? Describe their odors.

b. Do they have the same appearance (physical state, color, etc.)?

c. Draw complete structures for *n*-hexane and cyclohexane, showing each bond as a dash.

d. Draw the molecular structures for *n*-decane and for a component of mineral oil.

e. List the compounds tested in order of increasing viscosity.

f. If viscosity is related to the difficulty molecules have in slipping past one another, why should it increase with chain length?

4. Isomers

a. Compare the physical states of the two dichlorobenzenes.

b. In your own words, explain why these dichlorides are isomers.

c. Are the odors of the isomers *n*-butyl alcohol and *t*-butyl alcohol identical?

d. Record the number of drops of each alcohol that dissolve in 1 mL of water:

n-butyl alcohol _____ drops; and *t*-butyl alcohol _____ drops

e. On the basis of your observations on odors and solubilities, do these isomeric alcohols have different physical properties?

f. Which alcohol reacts more rapidly with sodium?

g. Write equations for the reactions. (Hint: recall the reaction between water and sodium, and note the similarity in molecular structure between an alcohol and water.)

h. Indicate (circle) how the odors of diethyl ether and the alcohols compare: very similar, somewhat similar, very different.

i. Do the compounds react (burn) differently with oxygen? Describe the burning characteristics of each.

j. Draw complete structures for the two alcohols and diethyl ether.

k. In your structures, circle the group of atoms (functional group) that characterize each compound as a member of its particular family.

The Hydrocarbons | 19

OBJECTIVES

1. To examine the physical characteristics of hydrocarbons and to recognize some of the hydrocarbons of everyday life.

2. To become familiar with reactions that are characteristic of each of the four principal types of hydrocarbons.

3. To learn to synthesize acetylene, the most important of the alkyne family of hydrocarbons.

DISCUSSION

Hydrocarbons are organic compounds that contain only carbon and hydrogen. They are extremely important to our technological society because so many products are derived from them: fabrics, plastics, antifreezes, medicinals, anesthetics, insecticides, fertilizers, paints, cleaning solvents, explosives, and a host of other items. The two major sources of hydrocarbons are petroleum (and natural gas) and coal tar. Petroleum is an extremely complex mixture of compounds, mainly aliphatic hydrocarbons. The aromatic hydrocarbons are obtained primarily from a process called "catalytic reforming," which involves converting aliphatic hydrocarbons to benzene and other aromatic derivatives.

Each of us, on the average, directly or indirectly uses several tons of petroleum each year, mostly for fuel—more than our bodily intake of food, water, or oxygen. This fact is especially sobering when we recall that petroleum is the only one of these natural resources that is not being replenished. It is imperative that we develop other sources of energy, such as solar energy and nuclear fusion so that the hydrocarbons can be reserved for making petrochemicals.

Hydrocarbons are also directly responsible for much pollution of our environment. Incomplete combustion of hydrocarbons in automobile engines produces carbon monoxide and other components of smog. Spills of crude petroleum have contaminated our beaches and our fishing beds. Here is a classic example of a product that is both a blessing and a curse, objectionable but indispensable. In today's experiment you will observe the physical properties of several different hydrocarbons and hydrocarbon derivatives and study some of their important chemical reactions.

Types of Hydrocarbons

Aliphatic hydrocarbons and aromatic hydrocarbons are two broad classes of compounds. Aliphatic hydrocarbons are divided into the alkanes, the alkenes, and the alkynes. These families of hydrocarbons and the distinctions between them are shown in table 19.1.

TABLE 19.1　The Types of Hydrocarbons

Type	Names and Structures of Characteristic Functional Groups		Simplest Member
Alkane*	Carbon-carbon single bond	$-\overset{\|}{\underset{\|}{C}}-\overset{\|}{\underset{\|}{C}}-$	Methane[†], CH_4 Ethane, CH_3-CH_3
Alkene (olefin)*	Carbon-carbon double bond	$\geq\!C\!=\!C\!\leq$	Ethylene, $CH_2\!=\!CH_2$
Alkyne*	Carbon-carbon triple bond	$-C\!\equiv\!C-$	Acetylene, $CH\!\equiv\!CH$
Aromatic	Benzene ring		Benzene,

*Aliphatic hydrocarbons.

[†]Methane is an alkane although it does not contain a carbon-carbon single bond.

Both chain and cyclic (ring) compounds are known for the three types of aliphatic hydrocarbons.[1]

$CH_3(CH_2)_4CH_3$　　$\underset{\text{Cyclohexane}}{\overset{\displaystyle CH_2\overset{CH_2-CH_2}{\diagup}\quad\diagdown\;CH_2}{\underset{CH_2-CH_2}{}}}$　　$\underset{\text{1-Hexene}}{CH_2\!=\!CH(CH_2)_3CH_3}$　　$\underset{\text{Cyclohexene}}{\overset{\displaystyle CH_2\overset{CH=CH}{\diagup}\quad\diagdown\;CH_2}{\underset{CH_2-CH_2}{}}}$

n-Hexane　　Cyclohexane

Hydrocarbons may be saturated or unsaturated. All those with carbon-carbon double or triple bonds (that is, all but alkanes) are unsaturated.

[1]Cycloalkynes with fewer than eight carbons in the ring are too unstable to exist.

Reaction with Bromine

In today's experiment you will use specific chemical reactions and other simple tests to distinguish between alkanes, alkenes, and aromatics. For example, bromine reacts differently with each of the three types of hydrocarbons,[2] as illustrated in the following equations.[3]

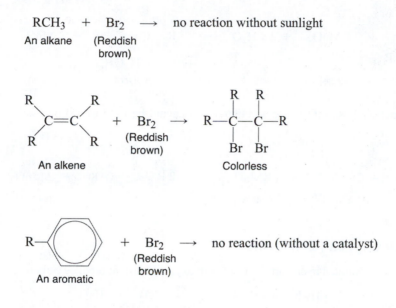

Whenever a reaction occurs, reddish brown bromine disappears and colorless products (alkyl bromides) are formed.[4] Conditions for, and evidence of, reaction are summarized in table 19.2.

TABLE 19.2 Reaction of Bromine with Alkanes, Alkenes, and Aromatics

Type of Hydrocarbon	Conditions for Reaction	Evidences of Reaction
Alkanes	No reaction without sunlight	
Alkenes	Reactions occur in the dark or in the light	Reddish brown color disappears (usually rapidly); no HBr is formed
Aromatics*	No reaction unless catalyst (Fe^{3+}) is present	Reddish brown color disappears slowly; HBr is formed

*Aromatic hydrocarbons with alkyl groups (for example, toluene, ⬡—CH₃) react with bromine in the presence of sunlight.

---

[2] Alkynes also add bromine and cannot be distinguished from alkenes by this procedure without considerable difficulty.

$$R—C \equiv C—R + 2\ Br_2 \rightarrow R—CBr_2CBr_2—R$$
(Reddish brown)   (Colorless)

[3] In this experiment, the R–group represents a hydrogen, an alkyl group (methyl, ethyl, etc.), or an aryl group (a phenyl or naphthyl ring).

[4] Occasionally the reaction solution becomes faintly yellow due to by-product formation.

Reaction with Potassium Permanganate

Potassium permanganate, $KMnO_4$, reacts with alkenes but not with alkanes and aromatics. Evidence for a reaction is the disappearance of purple $KMnO_4$ and the formation of brown manganese dioxide, MnO_2, as shown in the equation.

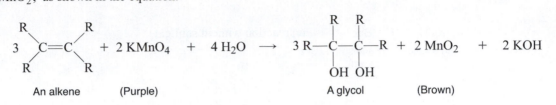

Alkynes also react with $KMnO_4$, and so this reagent cannot be used to distinguish between an alkyne and an alkene.

$$R—C\equiv C—R + 2\,KMnO_4 \longrightarrow 2\,R—\overset{\displaystyle O}{\overset{\|}{C}}—O^-K^+ + 2\,MnO_2$$

An alkyne   (Purple)   (Brown)

Reaction with Oxygen (Combustion)

Both saturated and unsaturated hydrocarbons react with oxygen (burn) to produce carbon dioxide and water.

$$\text{Hydrocarbon} + O_2 \rightarrow CO_2 + H_2O$$

However, most unsaturated hydrocarbons burn with a sooty flame. The quantity of soot is directly related to the degree of unsaturation of the hydrocarbon. Therefore, a sooty flame can serve as a test for unsaturation.

Synthesis of Acetylene

Most of the lower-molecular-weight chain hydrocarbons are obtained by "cracking" the large molecules found in crude petroleum. However, certain hydrocarbons, for example acetylene, can be synthesized easily in the laboratory. Today's experiment illustrates one method for its preparation.

Hydrocarbon Derivatives

Compounds that can be made from hydrocarbons are called hydrocarbon derivatives. The structures of three common hydrocarbon derivatives of benzene are

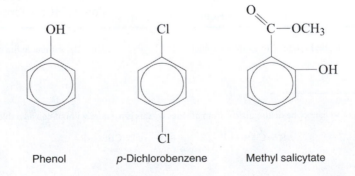

Phenol   *p*-Dichlorobenzene   Methyl salicytate

You will examine these compounds today.

ENVIRONMENT, CULTURE, AND CHEMISTRY
Hydrocarbons and Global Warming

Since the beginning of the Industrial Age, the temperature of the earth's surface has gradually warmed. This warming trend accelerated significantly in the 20th century, with the rate of increase essentially doubling during the last half of the century. This phenomenon has become known as Global Warming. The earth's temperature increased 0.75°F across the 20th century, and some scientists suggest that this value will increase fivefold by the end of the current century. The rise in temperature has roughly paralleled an increase in the concentration of "greenhouse gases." These gases, located in the atmosphere, consist primarily of water, carbon dioxide, and methane. They are called "greenhouse gases" because they trap the heat (infrared radiation), preventing it from escaping, much as glass panes in a greenhouse hold in the heat, and thus are believed to cause warming of the earth's surface. Carbon dioxide, produced primarily from combustion of hydrocarbons and coal, is considered the principal culprit in the temperature increase. About 3.5 billion metric tons per year of carbon, mainly as carbon dioxide, has been produced globally in recent years from the use of hydrocarbon fuels in automobiles, planes, other vehicles, and heating buildings; burning of coal in the production of electricity also adds significantly to the amount of carbon dioxide. This represents a 36% increase in the atmospheric concentration of carbon dioxide since 1750. The oceans hold many times this amount of carbon dioxide in solution, but the dissolving process is slow. There have been serious proposals to sequester (pump) liquid carbon dioxide into the ocean for storage. Although the increase in the global temperature is probably caused by the increase in "greenhouse gases," other explanations have been offered. It is an active area of debate among the public and even among some scientists.

Name _____

Date _____ Lab Section _____

Pre-Laboratory Questions | 19

1. Define the terms *hydrocarbon derivative, aliphatic hydrocarbon, unsaturation,* and *phenyl.*

2. What is the evidence for the reaction of potassium permanganate with alkenes? with alkynes?

3. What feature in its structural formula indicates whether a hydrocarbon is an alkane, alkene, alkyne, or aromatic compound?

4. Draw the electronic structure for benzene, showing valence electrons.

5. Write equations to illustrate the reaction of bromine with these organic compounds: ethane, C_2H_6; ethylene, C_2H_4; and acetylene, C_2H_2.

6. Can the combustion test be used to distinguish between an alkene and an alkyne? Explain.

EXPERIMENT

1. Physical Properties of Some Hydrocarbons

Place about 10 drops of cyclohexene, cyclohexane, and toluene[5] in three separate, clean dry test tubes. (To dry the test tubes simply use a twisted piece of paper towel.) Waft and compare the odors of these hydrocarbons. (Answer Question 1, a–b on the Observations and Results sheet.) **Caution:** *Never take more than a few drops of any hydrocarbon to your desk at one time because hydrocarbons are extremely flammable.*

Add about 10 drops of water to each test tube. Shake the test tubes and note whether the hydrocarbon dissolves. Repeat the solubility test in methylene chloride, CH_2Cl_2, with fresh samples of the hydrocarbons in clean test tubes. Do the hydrocarbons dissolve in CH_2Cl_2? (Answer Question 1, c–d on the Observations and Results sheet.) Compare the odor of each cyclic hydrocarbon with that of paraffin wax. (Answer Question 1, e–f.)

2. Reactions and Tests to Distinguish among the Hydrocarbons

A.   Reaction with Bromine

Place 10 drops each of cyclohexane and cyclohexene in separate clean, dry test tubes, and to each add 2 drops of dilute (5%) bromine in methylene chloride.[6] Observe whether a reaction (bromination) occurs. (Answer Question 2A, 1.)

B.   Reaction with Potassium Permanganate

Determine if the three cyclic hydrocarbons react with potassium permanganate, $KMnO_4$. Dissolve 6 drops of each hydrocarbon in 2 mL of ethyl alcohol (a solvent) in separate test tubes, and add 2 drops of 2% $KMnO_4$ solution. Is the purple color of the $KMnO_4$ solution replaced by a brown precipitate? For a positive test, a reaction should occur in 1 minute or less. The solvent, ethyl alcohol, reacts slowly with $KMnO_4$, producing a brown color in about 5 minutes. (Answer Question 2B, 1.)

C.   Reaction with Oxygen (Combustion)

Put a couple of drops of each of the three cyclic hydrocarbons, one at a time, on your spatula and ignite the hydrocarbons with your Bunsen burner (add the drops to the **cold** spatula with a dropping pipet, quickly touch the spatula to the flame, and remove it for observation while the hydrocarbon burns). Does the compound burn with a clean or sooty flame? (Do not confuse white vapor with soot.) (Answer Question 2C, 1–2.)

---

[5]See the footnote to table 19.2 for the structure of toluene. Benzene, the simplest aromatic hydrocarbon, is not used because of its toxicity.

[6]The bromine is diluted in CH_2Cl_2 because of the difficulty and danger of handling the pure liquid.

D. Tests on Some Common Hydrocarbons

Perform the three tests for unsaturation (Br_2, $KMnO_4$, and combustion) on the following common hydrocarbons: gasoline, turpentine, naphthalene, paraffin wax, polyethylene, and polystyrene. Use 6 drops of liquid and about 0.15 g (small pieces) of solid. Perform the $KMnO_4$ test on polyethylene and polystyrene by adding the $KMnO_4$ solution directly to the solid in a test tube; do not put either of them in alcohol solution. (Record your answers in table 2D, 1 and answer Question 2D, 2.)

> **Dispose of all waste from parts 1 and 2 in the appropriate containers.**

3. Observations on Some Common Hydrocarbon Derivatives

On the reagent shelf you will find the three common hydrocarbon derivatives whose structures and names are given in the discussion section: methyl salicylate, phenol, and *p*-dichlorobenzene. Waft and describe their odors. (Answer Question 3, a–b.)

4. Synthesis of Acetylene

Place a small piece (the size of a pea) of calcium carbide, CaC_2, in a 20- or 25-cm test tube and add 10 drops of water. After about 30 seconds, thrust a burning splint into the test tube. (Answer Question 4, a–e.)

To confirm by a second method that acetylene is formed in the reaction of water and calcium carbide and that it is unsaturated, prepare a filter-paper stick by folding a piece of filter paper over and over until it is a stick 5 to 7-cm long and 6-mm wide (figure 19.1). Then clean and *dry* a large test tube and drop in another piece of calcium carbide. Place a *tiny* drop of potassium permanganate solution on the end of the filter-paper stick. Add 10 drops of water to the calcium carbide in the test tube, and put your hand lightly over the test tube for about 30 seconds to compress the acetylene. Now quickly thrust the filter-paper stick into the test tube *without letting it touch the liquid,* and observe what happens to the purple color of the permanganate. Allow 2 or 3 minutes for the reaction to occur. (Answer Question 4, f.)

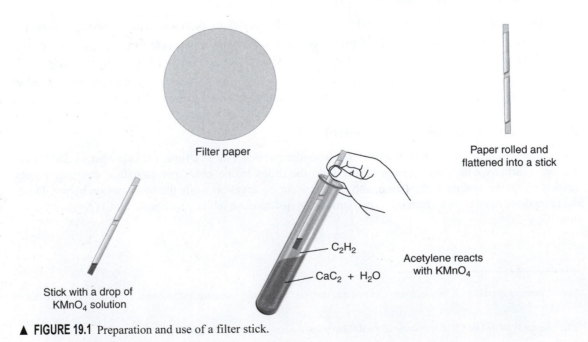

Filter paper

Paper rolled and flattened into a stick

Stick with a drop of $KMnO_4$ solution

C_2H_2

CaC_2 + H_2O

Acetylene reacts with $KMnO_4$

▲ **FIGURE 19.1** Preparation and use of a filter stick.

QUESTIONS AND PROBLEMS

1. Homemade jelly used to be preserved by placing a coating of paraffin over the surface. From a chemical standpoint, why is the hydrocarbon mixture called paraffin wax ideally suited for a covering on a jelly jar?

2. When hydrocarbons burn in insufficient oxygen, one product (along with water) is carbon monoxide, CO. Write an equation for the combustion of acetylene with insufficient oxygen. Why is some CO formed during combustion in an automobile engine?

3. Two of the compounds in this experiment had similar odors, yet their molecular structures are very different. Comment on factors at the molecular level that may be involved in the mechanism of smell. (Assume that a nerve impulse to the olfactory center of the brain is initiated by interaction of molecules, in this case the hydrocarbons, with a receptor in the nose.)

4. Discuss the relationship between the molecular weights of hydrocarbon molecules and the uses of hydrocarbons in modern society.

QUESTIONS AND PROBLEMS

EXPERIMENT

The Hydrocarbons | 19

OBSERVATIONS AND RESULTS

1. Physical Properties of Some Hydrocarbons

a. Draw the structures of the three cyclic hydrocarbons and indicate which is an alkane, which an alkene, and which an aromatic.

b. Compare the odors of the hydrocarbons. Did you notice an odor trend among the types of hydrocarbons? Comment.

c. Were the hydrocarbons soluble in water? in methylene chloride? What is your evidence that they dissolve in one solvent and not in the other?

d. Explain why they dissolve in one solvent, but not in the other.

e. Paraffin wax is primarily a mixture of long-chain hydrocarbons, an example of which is $C_{25}H_{52}$. Draw the structure of a molecule found in this mixture.

f. How do the odors of the cyclic hydrocarbons and paraffin compare? How do you explain the tremendous difference in intensity of the odors of paraffin and the cyclic hydrocarbons?

2. Reactions and Tests to Distinguish among the Hydrocarbons

A. *Reaction with Bromine*

1. Did bromine react immediately with cyclohexane? with cyclohexene? If either of the hydrocarbons reacted immediately with bromine, write an equation for the reaction.

B. *Reaction with Potassium Permanganate*

1. Which of the cyclic hydrocarbons reacted with potassium permanganate, $KMnO_4$? Write equations for the reaction(s) that occurred.

C. *Reaction with Oxygen (Combustion)*

1. Compare the amount of soot that was produced when the various cyclic hydrocarbons burned.

2. On the basis of your observations of the burning of toluene, comment on the practical problems associated with using toluene as fuel in your car.

D. *Tests on Some Common Hydrocarbons*

1. Record your observations on the tests of the common hydrocarbons with Br_2, $KMnO_4$, and O_2. Also indicate the type of hydrocarbon (alkane, alkene, or aromatic) in each case and whether it is saturated or unsaturated.

	Observation			Saturated or	Type		
Hydrocarbon	$KMnO_4$	Br_2	O_2	Unsaturated?	Alkane	Alkene	Aromatic
Gasoline	_____	_____	_____	_____	_____	_____	_____
Turpentine	_____	_____	_____	_____	_____	_____	_____
Napthalene	_____	_____	_____	_____	_____	_____	_____
Paraffin	_____	_____	_____	_____	_____	_____	_____
Polyethylene	_____	_____	_____	_____	_____	_____	_____
Polystyrene	_____	_____	_____	_____	_____	_____	_____

2. Using your textbook and any other source, draw structures for naphthalene, polystyrene, polyethylene, and a high-octane compound typically found in gasoline (show segments of polymer molecules).

3. Observations on Some Common Hydrocarbon Derivatives

a. Relate the odor of each of the hydrocarbon derivatives to some familiar commercial or natural material.

b. Which of the other hydrocarbons that you studied in this experiment has an odor similar to that of *p*-dichlorobenzene?

4. Synthesis of Acetylene

a. Describe what happened when water was added to calcium carbide.

b. What did you observe when the burning splint was thrust into the test tube?

c. Is acetylene unsaturated? What is your evidence?

d. Write an equation for the synthesis of acetylene.

e. Write an equation for the combustion of acetylene.

f. Did the purple color of potassium permanganate disappear in the presence of acetylene vapor? What color product was formed? Does this confirm that acetylene is unsaturated? Write an equation for the reaction that occurred.

Alcohols and Phenols | 20

OBJECTIVES

1. To learn about the physical properties of representative alcohols and phenols.
2. To examine reactions of alcohols that are of particular importance in biochemistry and physiology.
3. To become acquainted with some alcohols and phenols that are a part of our everyday life.

DISCUSSION

Alcohols

Methanol or methyl alcohol, CH_3OH, is the simplest member of the alcohol family, but it is not nearly as well known as ethanol, CH_3CH_2OH, the depressant drug in alcoholic beverages. Ethanol (often called ethyl alcohol) is also used as an antiseptic, a solvent for medicines and chemicals, and a preservative. Methyl alcohol may be viewed as the parent alcohol. If one of the hydrogens in the methyl group of methyl alcohol (a primary alcohol) is replaced by a carbon group (alkyl or aromatic), the resulting alcohol is still primary; if two hydrogens are replaced, a secondary alcohol is formed; and if all of the methyl hydrogens are replaced, a tertiary alcohol results. R represents the carbon group.

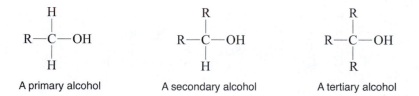

Ethylene glycol and glycerol (glycerine) are examples of alcohols that contain more than one hydroxyl group, —OH.

$$
\begin{array}{cc}
CH_2-CH_2 & CH_2-CH-CH_2 \\
| \quad\quad | & | \quad\quad | \quad\quad | \\
OH \quad OH & OH \quad OH \quad OH \\
\text{Ethylene glycol} & \text{Glycerol}
\end{array}
$$

Ethylene glycol is the principal component of antifreeze, and glycerol is produced in digestion and saponification of fats and oils.

Oxidation of Alcohols

Now let us consider some of the most important reactions of alcohols. Oxidation of primary and secondary alcohols gives aldehydes[1] and ketones, respectively. A common oxidizing agent is chromic anhydride, CrO_3, in sulfuric acid,[2] a mild oxidizing agent that will not oxidize tertiary alcohols.

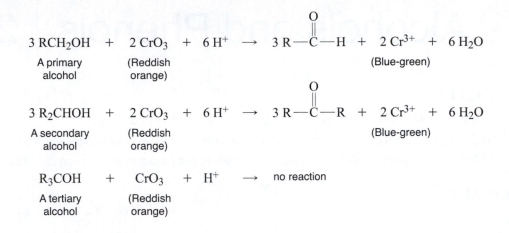

As indicated in the equations, a color change from the reddish orange of $CrO_3-H_2SO_4$ to the blue-green of Cr^{3+} confirms oxidation of an alcohol. Chromium is reduced from an oxidation state of 6+ in the anhydride to 3+ in Cr^{3+}. You will use the oxidation reactions in today's experiment to distinguish tertiary alcohols from primary and secondary alcohols.

Dehydration of Alcohols

Alcohols can be dehydrated (lose water) to alkenes (olefins) by heating them with a dehydrating agent such as H_2SO_4.[3]

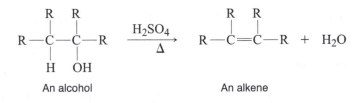

Dehydration reactions occur with all three types of alcohols: the alcohol loses an —OH group and a hydrogen located on adjacent carbons. These adjacent carbons then form a double bond to produce the alkene product. You will carry out a dehydration reaction today. Oxidation and dehydration reactions of

---

[1]Aldehydes are oxidized further to carboxylic acids unless the aldehyde is removed from the water mixture as it is formed (by distillation, for example).

[2]Net ionic equations are shown. The balanced molecular equation for the oxidation of a primary alcohol is

$$3\,RCH_2OH + 2\,CrO_3 + 3\,H_2SO_4 \rightarrow 3\,RCHO + Cr_2(SO_4)_3 + 6\,H_2O$$

[3]This is an example of intramolecular dehydration (dehydration occurring within a molecule). Competing intermolecular dehydration (dehydration involving two molecules) to give an ether also occurs.

$$2\,ROH \xrightarrow[\Delta]{H_2SO_4} 1\,ROR + H_2O$$

alcohols are very important in biochemistry. You will encounter many examples of both in your studies of the metabolism of carbohydrates and lipids and the synthesis of lipids.

Phenols

From the standpoint of structure, phenols and alcohols look much alike. Their chemical properties are, however, quite different. For example, does the structure of phenol, the simplest member of the family, permit dehydration to an alkene or oxidation to an aldehyde or ketone? The phenol family includes all compounds that have an —OH group bonded directly to an aromatic ring. Many phenols have strong odors that you may associate with disinfectants. Phenol itself is an important disinfectant for preserving biological specimens and is present in low concentrations in Chloraseptic, used to treat sore throats. A more complex phenol, methyl salicylate, is a main ingredient in Bengay, a treatment for sore muscles and in Listerine mouthwash. You have already become acquainted with methyl salicylate in Experiment 2.

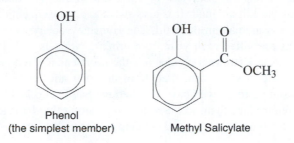

Phenol
(the simplest member)

Methyl Salicylate

Phenol in concentrated solution is highly toxic to all types of cells. If you spill phenol itself or certain other phenols on your skin, they will burn (kill) the tissue.[4] You will examine some common phenols in today's experiment.

Ferric Chloride Complex Formation with Phenols
(Distinguishing Test for Phenols)

Most phenols react with iron(III) here and in equation below ion, Fe^{3+}, in ferric chloride, $FeCl_3$, solution to form complexes varying greatly in color from greenish-gray to yellow-orange to deep purple. p-Chlorophenol is used to illustrate this reaction.

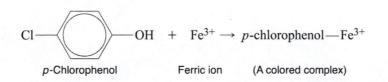

p-Chlorophenol   Ferric ion   (A colored complex)

You will use the formation of these complexes to detect the presence of phenols in some consumer products.

---

[4]Before the discovery of barbiturates, phenol was frequently taken as a poison to commit suicide. What a terrible death!

ENVIRONMENT, CULTURE, AND CHEMISTRY

Ethylene Glycol, a Useful but Dangerous Chemical

Ethylene glycol, a part of today's experiment, is sweet and odorless when pure. It is another example of a chemical that has many important (perhaps essential) functions in society, yet it has a "dark side" that makes it a real danger. It is used as antifreeze and in brake fluids, de-icing products, detergents, paints, cosmetics, shoe polish, inks, treatment for wood rot, preservation of biological specimens, and as a starting material for many polymers. Its major use (60%) is as antifreeze in engines, because ethylene glycol/water mixtures have a high boiling point and a low freezing point (as low as −45°F). Therefore, it protects an engine at low temperatures from freezing and cracking and does not boil away at high temperatures. Also, ethylene glycol itself, along with various additives, protects the engine and its components such as the radiator and water pump. What about the "dark side"? Ethylene glycol is one of the most toxic alcohols; as small an amount as 2 oz. (57 g) can kill an adult. It is also sweet and, consequently, it is attractive and lethal to children, pets, and animals in the wild. The Humane Society estimates that tens of thousands of dogs and cats are killed each year from drinking what is called by one writer, "sweet death." A few years ago, a California condor (on the short endangered species list), a majestic, soaring bird with a 9.5-foot wing span, died from drinking antifreeze that had leaked from a parked car. How devastating to all who have watched the San Diego Wild Animal Park's attempt to raise condors and reintroduce them to the wild. Considering the damage caused by antifreeze, the world would benefit from a less toxic and enticing (sweet) substitute.

Pre-Laboratory Questions | 20

1. Draw the structures of the functional groups that are characteristic of alcohols and phenols.

2. List the names and uses of the consumer products that you will study in today's experiment.

3. Define the terms *viscosity, oxidation, hydroxyl group,* and *hydrogen bonding.*

4. In this experiment, what is the evidence for the oxidation of an alcohol? for the dehydration of an alcohol?

EXPERIMENT

1. Alcohols

A. Physical Properties of Alcohols

Observe and waft the odors of the common alcohols ethyl alcohol (ethanol), isopropyl alcohol (2-propanol), and menthol. (Answer Question 1A, 1–2 on the Observations and Results sheet.)

B. Ethylene Glycol and Glycerol

Compare the viscosity of isopropyl alcohol, ethylene glycol, and glycerol by letting them drop slowly from a medicine dropper. (Answer Question 1B, 1.) Assemble an apparatus for determining boiling points (figure 20.1), using a dry, large (2.5-cm diameter) test tube, and a thermometer that reads to 250°C. Drop a boiling chip into the test tube, and add 1 mL of ethylene glycol. Adjust the bulb of the thermometer so that it is about 2.5 cm above the surface of the liquid; do not allow the bulb of the thermometer to touch the sides of the test tube. Heat the glycol with a low flame. As the liquid boils, you will see the line of condensed vapor creep up the side of the test tube. Continue to boil the glycol at such a rate that the condensation line remains for a few minutes about 2 cm above the bulb of the thermometer. When the temperature ceases to rise, you have reached the boiling point of ethylene glycol. (Record the temperature in 1B, 2 and answer Question 1B, 3–5.)

Dispose of any excess ethylene glycol in the appropriate container.

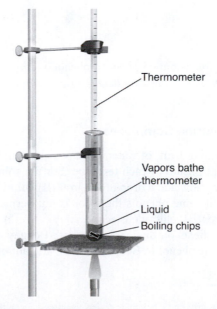

Thermometer

Vapors bathe
thermometer

Liquid

Boiling chips

▲ FIGURE 20.1 Apparatus for determining boiling points.

C. Oxidation of Alcohols

To each of three clean, dry test tubes add 1 mL of reagent acetone (acetone is a solvent). To the first test tube add 1 drop of *n*-butyl alcohol, $CH_3(CH_2)_3OH$; to the second add 1 drop of *sec*-butyl alcohol, $CH_3CH_2CH(OH)CH_3$; and to the third add 1 drop of *tert*-butyl alcohol, $(CH_3)_3COH$. Then to each test tube add 1 drop of chromic anhydride-sulfuric acid $(CrO_3—H_2SO_4)$ oxidizing agent, and mix each solution thoroughly. Oxidation is confirmed if the color of the solution changes from reddish orange to blue-green within a few minutes.

Now determine whether isopropyl alcohol and menthol (structure shown; small amount from a spatula) are oxidized by the $CrO_3—H_2SO_4$ reagent. Again use acetone as a solvent. (Answer Question 1C, 1–3.)

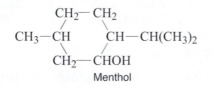

Menthol

D. Dehydration of an Alcohol[5]

For this experiment you will need a filter-paper stick (see figure 19.2), a small test tube mounted in a clamp on a ring stand, and a few drops of dilute (2%) potassium permanganate, $KMnO_4$, solution.

To the test tube add 10 drops of *tert*-butyl alcohol and 1 drop of concentrated H_2SO_4. **Caution:** *Concentrated H_2SO_4 is dangerous! If you come in contact with the acid, wash it off immediately*. Put a tiny drop of $KMnO_4$ solution on the end of the filter-paper stick, and hold the $KMnO_4$ about 2 or 3 cm above the surface of the alcohol. Now *heat* the alcohol with a Bunsen burner with a low flame until *it just begins to boil*. What happens to the color on the end of the filter-paper stick? (Answer Question 1D, 1–5.)

Do not let the $KMnO_4$ come into contact with the alcohol at any time.

2. Phenols

A. Physical Properties

Describe and waft the odors of three common phenols: phenol, methyl salicylate, and eugenol. (Answer Question 2A, 1–4.) **Caution:** *Do not allow the phenols to come in contact with your skin; they might cause serious burns.*

B. Formation of Iron(III) Chloride Complexes

To each of three test tubes containing 1 mL of water add a pea-sized amount of one of the three phenols that you just smelled—a different phenol to each test tube. Stir the solutions with a stirring rod to dissolve as much phenol as possible. Now add a drop of 1% iron(III) chloride solution and note the color of the complex that is formed. With some phenols the color is not permanent, so the solution should be watched closely the instant the iron(III) chloride is added.

Establish by reading the ingredients on the labels that both Bengay and Chloraseptic contain a phenol. To confirm the presence of the phenol in Bengay, mix an amount of Bengay the size of a large pea

---

[5]As a background for this experiment, you may find it helpful to review the discussion on the reactions of alkenes with potassium permanganate in Experiment 19.

in 1 mL of water in a test tube. Stir in 5 drops of $FeCl_3$ and observe the color. To test for the phenol in Chloraseptic, mix 10 drops of Chloraseptic with 5 drops of $FeCl_3$ solution in a test tube. Note the color change. (Since Chloraseptic is already colored, formation of a complex is indicated by a change in the color of the solution.) (Answer Question 2B, 1–4.)

Dispose of all waste in the appropriate container.

QUESTIONS AND PROBLEMS

1. Is the following alcohol primary, secondary, or tertiary? Is it aromatic?

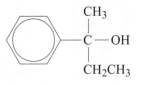

   Would CrO_3—H_2SO_4 oxidize it to an aldehyde or a ketone? If so, write an equation for the reaction. Would it undergo dehydration with sulfuric acid? If so, write the structures of the products.

2. Write equations for reactions that could be used to distinguish between the isomeric alcohols.

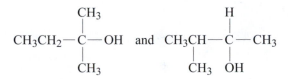

3. Write structures for all the products that could be formed if menthol were dehydrated (intramolecularly) with sulfuric acid.

4. Write an equation for the oxidation of isopentyl mercaptan (a component of skunk scent) with potassium bromate.

5. The structure of the amino acid tyrosine is

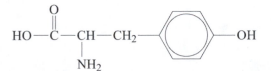

   Is tyrosine an alcohol? A phenol? On the basis of today's experiment how would you prove whether it is a phenol or an alcohol?

6. List at least three important uses for glycerol. Use your textbook or another resource book to answer this question.

Name _____

Date _____ Lab Section _____

Alcohols and Phenols

OBSERVATIONS AND RESULTS

1. Alcohols

A. *Physical Properties of Alcohols*

1. Which alcohol of those tested has the most pleasant odor? Which has the least pleasant odor?

2. Draw the structure and give an important use for each of the alcohols.

   Methyl alcohol:

   Ethyl alcohol:

   Menthol:

   Isopropyl alcohol:

B. *Ethylene Glycol and Glycerol*

1. Arrange the alcohols that you tested in the order of increasing viscosity, and use the concept of hydrogen bonding to explain the difference in viscosity.

2. Observed boiling point of ethylene glycol   _____ °C
3. Reported boiling point of ethylene glycol   _____ °C

4. The molecular weights of isopropyl alcohol and ethylene glycol are very close, 60 and 62, respectively, but their boiling points are far apart (the boiling point of isopropyl alcohol is 82°C). Explain.

5. Name at least two properties of ethylene glycol that make it an ideal automobile antifreeze.

C. *Oxidation of Alcohols*

1. Indicate which alcohols were oxidized: *n*-butyl alcohol, *sec*-butyl alcohol, *tert*-butyl alcohol, isopropyl alcohol, and menthol. Write equations for two oxidation reactions:

2. Which of the alcohols were oxidized to aldehydes?

3. Which of the alcohols are secondary alcohols?

D. *Dehydration of an Alcohol*

1. Did the potassium permanganate, $KMnO_4$, test prove that dehydration of *tert*-butyl alcohol occurred to give an alkene? What is the evidence?

2. Give the name and structure of the alkene.

3. Look up and record the boiling point of the alkene. Is it a gas at room temperature?

4. Write an equation for the formation of the alkene.

5. Write an equation for the reaction that occurred between the alkene and $KMnO_4$.

2. Phenols

A. *Physical Properties*

1. Describe the odor of phenol. Have you smelled it before? Where?

2. Methyl salicylate is also called oil of wintergreen because it is found in wintergreen berries. Where have you smelled methyl salicylate before? Give two sources in addition to Bengay, which is part of this experiment.

3. Eugenol occurs in a common spice. Name the spice and list other uses, particularly medicinal, for eugenol.

4. The structure of eugenol is

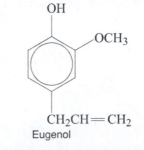

Eugenol

Identify the functional groups that are present in eugenol.

B. *Formation of Iron(III) Chloride Complexes*

1. Give the color of the complexes for phenol, methyl salicylate, and eugenol.

2. From the contents on the label, give the names, structures, and amounts (percentages) of the phenols in Bengay and Chloraseptic sore throat spray.
   Bengay:

   Chloraseptic sore throat spray:

3. Does the iron(III) chloride test confirm the presence of the phenol in the consumer products? Explain.

4. What is the function of the phenol in each of these products?

Aldehydes and Ketones | 21

OBJECTIVES

1. To investigate the physical properties and uses of some common aldehydes and ketones.

2. To develop familiarity with reactions of aldehydes that are of particular importance in biochemistry.

3. To learn to use characteristic reactions to distinguish between aldehydes and ketones, and between aliphatic and aromatic aldehydes.

DISCUSSION

Formaldehyde is a sharp, irritating gas and is the simplest member of the aldehyde family. Acetone, the simplest ketone, is on the opposite end of the odor spectrum; it is responsible for the pleasant odor of fingernail-polish remover.

Formaldehyde   Acetone

Both aldehydes and ketones contain the carbonyl group. Ketones have two R–groups[1] attached to the carbonyl carbon. In the case of aldehydes, at least one of the attached groups is a hydrogen.

Carbonyl group   An aldehyde   A ketone

---

[1]In this experiment R– denotes the alkyl groups methyl, CH_3-; ethyl, CH_3CH_2-; and benzyl, $C_6H_5CH_2-$. It also denotes the aromatic (aryl) groups phenyl, C_6H_5-, and naphthyl, $C_{10}H_7-$.

An aromatic aldehyde (or ketone) is one in which an aromatic ring (an R–group) is bonded directly to the carbonyl carbon. Benzaldehyde is the simplest aromatic aldehyde.

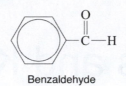

Benzaldehyde

In this experiment you will be introduced to several common aldehydes and ketones through their odors and other physical properties. Many of them have unique and interesting uses, which you will learn about. In general, aldehydes are most useful in the synthesis of other organic compounds. Ketones are used mainly as solvents.

Oxidation of Aldehydes: Distinction from Ketones

Aldehydes are easily oxidized to carboxylic acid, but ketones are not oxidized at all under ordinary reaction conditions. This difference serves as a convenient way to distinguish between aldehydes and ketones.

$$R-\overset{\overset{O}{\|}}{C}-H + (O) \rightarrow R-\overset{\overset{O}{\|}}{C}-OH$$

$$R-\overset{\overset{O}{\|}}{C}-R + (O) \rightarrow \text{no reaction}$$

The oxidation can be accomplished by mild oxidizing agents such as cupric ion, Cu^{2+}, in basic solution (Fehling's reagent),[2] or by silver ion, Ag^+, in ammonia solution (Tollens' reagent). Net ionic reactions[3] of an aldehyde with Fehling's reagent and with Tollens' reagent, respectively, are

---

[2]Often Benedict's reagent is used instead of Fehling's. Both reagents contain cupric ion, Cu^{2+}, in basic solution. Since cupric hydroxide is insoluble, special stabilizing agents are added to prevent precipitation. The stabilizing agent in Benedict's reagent is sodium citrate, and in Fehling's reagent it is sodium tartrate. The use of Fehling's reagent is recommended.

[3]Under the basic conditions of the Fehling's and Tollens' tests the salts of carboxylic acids are formed. For convenience, they are represented as acids in the equations. The complete balanced molecular equations for the reactions of aldehydes with Fehling's and Tollens' reagents are

$$R-\overset{\overset{H}{|}}{C}=O + 2\,Cu(OH)_2 + NaOH \rightarrow R-\overset{\overset{O}{\|}}{C}-O^-Na^+ + Cu_2O + 3\,H_2O$$

$$R-\overset{\overset{H}{|}}{C}=O + 2\,Ag(NH_3)_2^+ + 2\,OH^- \rightarrow R-\overset{\overset{O}{\|}}{C}-O^-NH_4^+ + 2\,Ag + H_2O + 3\,NH_3$$

$$R-\overset{\overset{\displaystyle O}{\|}}{C}-H \ + \ Cu^{2+} \ \xrightarrow{NaOH, \ H_2O} \ R-\overset{\overset{\displaystyle O}{\|}}{C}-OH \ + \ Cu_2O$$

(Deep blue)   Brick-red

$$R-\overset{\overset{\displaystyle O}{\|}}{C}-H \ + \ Ag^+ \ \xrightarrow{NH_3, \ H_2O} \ R-\overset{\overset{\displaystyle O}{\|}}{C}-OH \ + \ Ag$$

A positive Fehling's test is indicated by the formation of a brick-red precipitate, which in a blue solution may appear to be greenish. A positive Tollens' test is confirmed by the formation of a silver mirror on the inside of the test tube. Aromatic aldehydes do not give a positive Fehling's test, so this test can be used to distinguish between aromatic and aliphatic aldehydes. Both types of aldehydes react with Tollens' reagent. The Fehling test for aldehydes is used extensively in carbohydrate chemistry. For example, the simple sugar glucose is an aldehyde and is readily oxidized by Fehling's reagent. You will study these reactions in Experiment 26.

Formation and Hydrolysis of Acetals

An important reaction of aldehydes, particularly as it applies to carbohydrate chemistry, is the reversible formation of acetals. Acetals can be made by heating aldehydes with alcohols in the presence of a trace of acid. Hemiacetals are involved as intermediates in these reactions. In most cases the equilibrium lies to the left, in favor of the reactants, aldehyde and alcohol.

The functional group of an acetal is characterized by a central carbon to which are bonded two alkoxy groups (—OR′), a hydrogen, and an R–group; the acetal of formaldehyde is unique in that the central carbon bears two hydrogens (the R–group is hydrogen).

$$R-\overset{\overset{\displaystyle H}{|}}{\underset{\underset{\displaystyle OR'}{|}}{C}}-OR' \qquad\qquad H-\overset{\overset{\displaystyle H}{|}}{\underset{\underset{\displaystyle OR'}{|}}{C}}-OR'$$

General formula for acetals   General formula for acetals of formaldehyde

Hemiacetals are usually unstable and cannot be isolated. Glucose is an exception, existing predominately as a hemiacetal both in the solid state and in aqueous solution. Since the formation of an acetal is reversible, an aldehyde and an alcohol should be produced when an acetal is warmed in aqueous acidic solution (hydrolysis). You will study the hydrolysis of an acetal in today's experiment and later in experiments 26 and 27.

Iodoform Reaction of Methyl Ketones

Methyl ketones (ketones that have a methyl group bonded directly to the carbonyl group) react with iodine in basic solution to give iodoform, CHI_3, and other products.[4]

$$R-\overset{\overset{\textstyle O}{\|}}{C}-CH_3 \ + \ 3\,I_2 \ + \ 4\,NaOH \ \longrightarrow \ CHI_3 \ + \ R-\overset{\overset{\textstyle O}{\|}}{C}-O^-Na^+ \ + \ 3\,NaI \ + \ 3\,H_2O$$

A methyl ketone   Iodoform
(a yellow solid)

Iodoform is a yellow solid with a strong medicinal odor; it is used as an antiseptic. This reaction, known as the iodoform test, can be used to distinguish between methyl ketones and other ketones.

ENVIRONMENT, CULTURE, AND CHEMISTRY
Formaldehyde, A Simple Yet Complex Aldehyde

Formaldehyde is used with other organic monomers (molecules that condense to polymers) to make mixed polymers such as urea-formaldehyde resin, melamine resin, phenol formaldehyde resin, and many others. These mixed formaldehyde resins (polymers) are found in insulation in homes and vehicles, many other automobile parts such as brake shoes and pads, glues, adhesives in plywood and carpeting, topical creams and cosmetics, facial tissue, table napkins, and paper towels. It is fair to say that formaldehyde-containing polymers are spread broadly across industry and into everyday life. There is one serious drawback: most formaldehyde polymers are known to break down, releasing small amounts of formaldehyde, a known carcinogen (causes cancer). This breakdown becomes most serious in closed spaces and when the space is warmed. The Center for Disease Control (CDC) has set a limit of 8 ppb (parts per billion in air) above which formaldehyde is considered dangerous. In 2006, the trailers and mobile homes taken to New Orleans for people displaced by Hurricane Katrina, were found to contain 77 ppb—far exceeding the CDC level. Some of the people living in these trailers experienced nosebleeds, difficulty in breathing, and headaches. Since formaldehyde polymers are present in materials in the interior of automobiles, it is almost certain that the CDC level is exceeded when driving for extended lengths of time without air circulation. A level of 0.1 ppm (parts per million—nearly one thousand times greater than the CDC value) of formaldehyde causes the eyes to burn and makes breathing difficult for many people. Formaldehyde in water, a solution called formalin, was used for many years to preserve biological specimens, but was discontinued because of concern about cancer and eye irritation from the formaldehyde gas that evaporated. Can anything be done about the formaldehyde-polymer problem? One approach would be to make a real effort not to use formaldehyde resins in interior spaces. This cannot be done suddenly but steps can be taken in this direction. Finally, one other curious twist in the world of formaldehyde: many on-line comments state that formaldehyde is formed from aspartame, the sweetener called Nutrasweet (also sold under other brand names), leading to all types of illnesses. The claim is that the methoxy group (CH_3O) from aspartame is removed and converted to methyl alcohol, followed by oxidation to formaldehyde. Several thorough studies by chemists have established that this does not occur.

---

[4]The following compounds also give a positive iodoform test: all secondary alcohols, which are oxidized to methyl ketones by iodine; acetaldehyde; and ethyl alcohol, which is oxidized to acetaldehyde.

Pre-Laboratory Questions | 21

1. Consider the following aldehydes and ketones, which you will study in today's experiment.

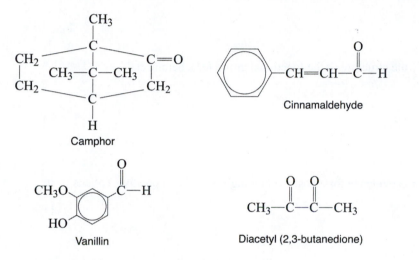

Camphor

Cinnamaldehyde

Vanillin

Diacetyl (2,3-butanedione)

   a. Which of the compounds are aldehydes? Which are ketones?

   b. Which of the aldehydes are aliphatic? Which are aromatic?

   c. Which compounds would react with Fehling's reagent? With Tollens' reagent? Write equations for all reactions that occur.

d. Draw the structural formula of an acetal, formed from one of the aldehydes (your choice), ethyl alcohol, and a trace of acid.

e. Which of the compounds would give a positive iodoform test? Write equations for reactions that occur.

2. Draw the structure of the functional group for a hemiacetal.

3. What is evidence for a positive Fehling's test? For a positive Tollens' test?

EXPERIMENT

1. Physical Properties

A. Odors and Uses

The following common aldehydes and ketones are on the reagent shelf: benzaldehyde, vanillin, camphor, cinnamaldehyde, diacetyl (2,3-butanedione). Waft each of the above compounds and note the odor of each of them (answer Question 1A, 1–14 on the Observations and Results sheet). In the first pre-laboratory question you are asked to classify some of these compounds as aldehydes or ketones. If you are still uncertain, ask your instructor.

B. Acetone as a Solvent

Paint some colored fingernail polish on the bottom of a dry test tube and allow the polish to dry. When it is dry, find out if acetone is a good solvent for this polymer. Put some acetone on a piece of paper towel and rub the dry polish with it. Commercial fingernail-polish remover is on the reagent shelf. Examine it (answer Question 1B, 1–3).

2. The Oxidation of Aldehydes: Fehling's and Tollens' Tests

A. Fehling's Test

To each of three test tubes add 2 mL of Fehling's reagent, freshly prepared by mixing equal amounts of solutions A and B, or add 2 mL of Benedict's reagent. To the first test tube add 3 drops of formaldehyde (formalin) solution,[5] to the second add 3 drops of benzaldehyde, and to the last add 3 drops of reagent-grade acetone. Stir the solutions thoroughly, and place them in a beaker of boiling water for no more than 10 minutes (figure 21.1).[6] (Answer Question 2A, 1–3.) *Formation of trace amounts of Cu_2O precipitate is not a confirming test.*

> **Dispose of all waste in the appropriate container.**

B. Tollens' Test

Prepare the Tollens' reagent as follows. To each of three clean test tubes add 1 mL of a dilute (5%) aqueous silver nitrate, $AgNO_3$, solution, and 1 drop of 5% sodium hydroxide solution. To each test tube add 1 mL of dilute (2%) ammonium hydroxide, and then *stir* in additional ammonium hydroxide drop by drop until only a few particles of silver oxide remain. *Avoid an excess of ammonium hydroxide.* Now conduct the Tollens' test on the following aldehydes and ketones. To the first test tube add 2 drops of formaldehyde solution, to the second add 2 drops of benzaldehyde, and to the third add 2 drops of reagent-grade acetone. The formaldehyde and acetone reactions should be done at room temperature; the

---

[5]The aqueous solution is called formalin. Pure formaldehyde is a gas, some of which escapes from water.

[6]Use a hot plate if your instructor so directs.

▲ **FIGURE 21.1** Fehling's test. A brick-red precipitate indicates the presence of an aldehyde.

benzaldehyde reaction must be carried out in the boiling water bath. Stir the solutions thoroughly and let them stand for 10 to 15 minutes. Does a Tollens' silver mirror form? (Answer Question 2B, 1–3.)

> **Caution:** Discard the Tollens' test reagents as directed when you have completed this part of the experiment. Explosive compounds may form if the solutions are allowed to stand. The silver mirror can be removed from the test tube by washing it with dilute nitric acid.

3. Hydrolysis of an Acetal

Dissolve 5 drops of methylal (dimethoxymethane), $CH_2(OCH_3)_2$, in 1 mL of 2% sulfuric acid solution. Stir the solution, and place it in a boiling water bath for 5 minutes. To the hydrolyzed solution in the test tube, add 2 mL of Fehling's reagent (equal amounts of solutions A and B),[7] and continue to heat the solution in the boiling water bath. Does the Fehling test confirm that an aldehyde is one of the products of hydrolysis of methylal? (Answer Questions 3, a–c.)

4. Test for a Methyl Ketone: Synthesis of an Antiseptic

Stir in a test tube 10 drops of 10% sodium hydroxide, 4 drops of acetone, and 1 mL of water. Then, while shaking the container, add drops of iodine-potassium iodide reagent *until the color of iodine disappears quite slowly.* (Approximately 5–10 drops of iodine-potassium iodide reagent will be required.) Does a precipitate form? Isolate the iodoform by pouring the reaction mixture onto a filter paper in a funnel. Dry the crystals between several pieces of filter paper. Compare the physical properties (odor and color) of your iodoform with those of the commercial iodoform on the reagent shelf. Place your iodoform on a piece of filter paper (be sure to write your name on the paper first) and give it to your instructor. (Answer Questions 4, a–b.)

> **Dispose of the filtrate and all waste from parts 3 and 4 in the appropriate containers.**

---

[7]Benedict's reagent does not work well in this part of the experiment.

QUESTIONS AND PROBLEMS

1. Methyl nonyl ketone is an active ingredient in certain commercial cat and dog repellents. Draw the structure for this ketone. Would this ketone give a positive iodoform test? If so, write an equation for this reaction.

2. Metaldehyde is used as a garden bait to attract and kill snails and slugs. It is synthesized by polymerizing four molecules of acetaldehyde into a cyclic tetramer.

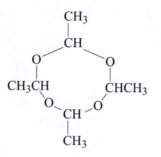

Draw a circle around each monomeric unit (acetaldehyde) in the metaldehyde molecule. Metaldehyde can be converted to acetaldehyde by treatment with acid. Write an equation for this reaction. Metaldehyde has a high vapor pressure. Why is this essential to its function as snail bait?

3. The ethyl alcohol that is consumed in beverages is converted to acetaldehyde in the cells. Acetaldehyde is quite toxic. On the basis of reactions that you studied in today's experiment, suggest how the cells detoxify (destroy by a reaction) acetaldehyde.

4. Chloral hydrate, $Cl_3CCH(OH)_2$, commonly known as knockout drops, is widely used as a sleep-producing (soporific) drug. Write an equation to show how chloral hydrate could be made from chloral, Cl_3CCHO. Would chloral react with Fehling's reagent? Would chloral hydrate react?

EXPERIMENT

Aldehydes and Ketones | 21

OBSERVATIONS AND RESULTS

1. Physical Properties

A. *Odors and Uses*

   1. Draw the structure for benzaldehyde and circle the aldehyde functional group.

   2. Describe the odor of benzaldehyde, which is a major component of oil of bitter almonds.

   3. On the basis of the odor, identify the commercial use for vanillin.

   4. Name all of the functional groups that are present in the vanillin molecule.

   5. Would vanillin give a positive iron(III) chloride test? (Refer to experiment 20.) Explain.

   6. Draw the structure of camphor and circle the carbonyl group.

7. Give the name of a compound, which you studied in an earlier experiment, that has an odor similar to camphor.

8. What is an important commercial use for camphor?

9. Draw the structure of cinnamaldehyde and circle the phenyl ring.

10. Cinnamaldehyde occurs in a very important household product. Name this product.

11. Would you expect cinnamaldehyde to react with bromine (Br_2)? If so, write an equation for this reaction.

12. Diacetyl is partially responsible for the aroma and taste of butter. Did you recognize the relationship between the chemical that you smelled and the odor of butter? Why is diacetyl called a diketone?

13. Draw the structure of formaldehyde.

14. Formaldehyde dissolves in water to form a hydrate. (Formation of a hydrate involves addition of water to the carbonyl group, analogous to the hydration of an alkene.) At equilibrium, the percentage of "free" aldehyde in a 30% "formaldehyde" solution at room temperature is only 0.01%. Write an equation that completely describes the hydration of formaldehyde.

B. *Acetone as a Solvent*

   1. Draw the structure for acetone and circle the alkyl groups.

   2. Does acetone dissolve fingernail polish (a polymer)?

   3. Examine the ingredients on the label of a container of commercial fingernail-polish remover. Is acetone present? Is the amount (percentage) of acetone indicated? If so, what is it?

2. The Oxidation of Aldehydes: Fehling's and Tollens' Tests

A. *Fehling's Test*

   1. Indicate which of the carbonyl compounds gave a positive test with Fehling's reagent: formaldehyde, benzaldehyde, and acetone.

   2. Write equations for the carbonyl compounds that reacted with Cu^{2+} in Fehling's reagent.

   3. Can Fehling's test be used to distinguish between an aldehyde and a ketone? Between an aromatic and aliphatic aldehyde? Explain.

B. *Tollens' Test*

   1. Indicate which of the carbonyl compounds gave a positive test with Tollens' reagent: formaldehyde, benzaldehyde, and acetone.

2. Write equations for the carbonyl compounds that reacted with Ag^+ in Tollens' reagent.

3. Can Tollens' reagent be used to distinguish between an aldehyde and a ketone? Between an aromatic aldehyde and an aliphatic aldehyde? Explain.

3. Hydrolysis of an Acetal

a. Write the structure for methylal and circle the acetal functional group.

b. What compounds were formed by the hydrolysis of methylal? Did Fehling's test confirm the presence of one of these compounds? Explain.

c. Write an equation for the hydrolysis of methylal, showing the structures of the intermediate hemiacetal and the products, and name all of the products.

4. Test for a Methyl Ketone: Synthesis of an Antiseptic

a. Describe what happened as you added the iodine solution to acetone.

b. Compare the color and odor of your iodoform with those of commercial iodoform.

Carboxylic Acids and Esters | 22

OBJECTIVES

1. To find out about the interesting roles that acids and esters play in daily life.
2. To examine several typical reactions of acids and esters, which will be of particular importance in biochemistry.

DISCUSSION

You are already familiar with one member of the carboxylic acid family, acetic acid, which is responsible for the sharp odor and sour taste of household vinegar. You may recall the "sour," disagreeable odor of rancid butter; butyric acid is the villain in this case. Goats owe their distinctly personal B.O. to the presence of a mixture of carboxylic acids, including valeric acid, in their urine. From a structural standpoint, the family characteristic of the carboxylic acids is the carboxyl group.

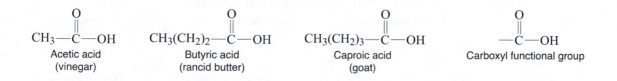

$$CH_3-\overset{\overset{\displaystyle O}{\|}}{C}-OH$$
Acetic acid
(vinegar)

$$CH_3(CH_2)_2-\overset{\overset{\displaystyle O}{\|}}{C}-OH$$
Butyric acid
(rancid butter)

$$CH_3(CH_2)_3-\overset{\overset{\displaystyle O}{\|}}{C}-OH$$
Caproic acid
(goat)

$$-\overset{\overset{\displaystyle O}{\|}}{C}-OH$$
Carboxyl functional group

The smaller aliphatic carboxylic acids are liquids, whereas the aromatic acids and dicarboxylic acids are solids. The sources of benzoic acid, an aromatic acid, and some dicarboxylic acids are

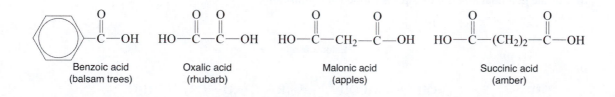

Benzoic acid
(balsam trees)

Oxalic acid
(rhubarb)

Malonic acid
(apples)

Succinic acid
(amber)

Several dicarboxylic acids, succinic acid among them, play a prominent role in the Krebs Acid Cycle, a metabolism cycle that you will study in biochemistry.

In contrast to carboxylic acids, esters have pleasant odors, many of which are associated with various fruits, flowers, and berries. Some esters, the related fruits, and the characteristic functional group are shown here.

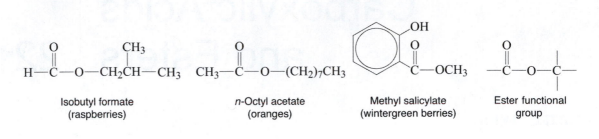

Isobutyl formate
(raspberries)

n-Octyl acetate
(oranges)

Methyl salicylate
(wintergreen berries)

Ester functional
group

Triglycerides (fats and oils) and waxes are important, naturally occurring esters. You will study triglycerides in a later experiment. Let us now consider some of the reactions of carboxylic acids and esters that you will investigate directly or indirectly in today's experiment.

Ionization of Carboxylic Acids

Carboxylic acids are weak acids and ionize only slightly in aqueous solution to give a hydrogen ion and a carboxylate anion (when R– is a hydrogen, alkyl, or aryl group).

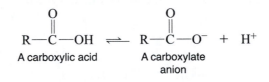

A carboxylic acid

A carboxylate
anion

At equilibrium, most of the acid is present in water as the un-ionized molecular acid. Only a few of the lower-molecular-weight acids (up to four carbons) are soluble in water.

Esterification and Hydrolysis

Esterification of a carboxylic acid by an alcohol to give an ester is reversible. Normally only 60% to 70% of the potential amount of ester is formed at equilibrium.

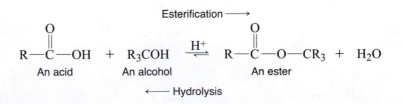

An acid   An alcohol   An ester

The reverse reaction, the reaction of an ester with water to give an alcohol and an acid, is called *hydrolysis*. The equilibrium can be shifted to the right to give more ester by the addition of excess acid or alcohol (normally the latter), or by removal of water. Hydrolysis is favored by an excess of water. Both reactions are catalyzed by inorganic acids.

Reactions with Sodium Hydroxide

Neutralization of Carboxylic Acids

Carboxylic acids react with bases, such as sodium hydroxide, to give carboxylate salts and water.

$$
\underset{}{R-\overset{\overset{\displaystyle O}{\|}}{C}-OH} \; + \; NaOH \; \longrightarrow \; \underset{\text{A carboxylate salt}}{R-\overset{\overset{\displaystyle O}{\|}}{C}-O^-Na^+} \; + \; H_2O
$$

The salts are usually soluble in water, whereas the acids frequently are not. Sodium carbonate, Na_2CO_3, and sodium bicarbonate, $NaHCO_3$, also neutralize acids, forming the carboxylate salts, carbon dioxide, and water.

Saponification of Esters

Hydrolysis of an ester in basic solution is called *saponification*. The products of the reaction are a salt and an alcohol.

$$
\underset{\text{An ester}}{R-\overset{\overset{\displaystyle O}{\|}}{C}-O-CR_3} \; + \; NaOH \; \longrightarrow \; \underset{\text{A carboxylate salt}}{R-\overset{\overset{\displaystyle O}{\|}}{C}-O^-Na^+} \; + \; \underset{\text{An alcohol}}{R_3COH}
$$

Saponification reactions can generally be "followed" visibly since most esters are insoluble in water, and the salts and low-molecular-weight alcohols are soluble. The ester layer gradually decreases as the reaction proceeds. Addition of dilute acid to the solution of a carboxylate salt converts it to a carboxylic acid.

$$
\underset{}{R-\overset{\overset{\displaystyle O}{\|}}{C}-O^-Na^+} \; + \; HCl \; \longrightarrow \; R-\overset{\overset{\displaystyle O}{\|}}{C}-OH \; + \; NaCl
$$

The carboxylic acid, RCOOH, is largely un-ionized, and, unless it is of low molecular weight, forms an insoluble layer or precipitates as a solid.

ENVIRONMENT, CULTURE, AND CHEMISTRY
Odors, Molecular Structures, and Organic Families

Smell is the most sensitive of our several senses, and two of the families of organic compounds with the most pronounced odors are the aliphatic carboxylic acids and their esters. No question, odors both significantly enhance and distract from our environment. Odorant molecules are also the most pronounced source of memory recall; one whiff of spearmint, the years roll away, and we are once again kids on the bank of a river in Southern Michigan where spearmint grows wild. Humans are capable of detecting certain odors to the incredibly low level of one part of compound in a trillion parts of air. Scientists believe that the differences in odors of various compounds are related to a combination of the molecular shape and electronic factors of the molecules as they bind to receptor sites in the odorant-binding proteins located at the top rear of the nasal cavity; this is perhaps best imagined as a type of lock and key relationship. When sufficient odorant molecules are bound to the receptor site, an electronic signal is sent to the brain and the odor/compound relationship is generated. Many families of organic compounds can be recognized by their characteristic odors. Aliphatic acids, RCO_2H, where the R group represents a saturated carbon chain, have sharp, disagreeable smells like those of butyric acid in rancid butter, and caproic acid in goat, and Limburger cheese. On the other hand, if the shape and electronic properties of the R group are changed substantially with a bulky, aromatic phenyl (benzene) ring, as in benzoic acid, the smell is quite pleasant. The delicious, fruity fragrance of the ester family (RCO_2R) seems to be less affected by shape. This is evident in the relatively small isobutyl formate molecule (smell of raspberries), and the much larger methyl salicylate molecule (oil of the wintergreen berry), which are both delightful to smell. Perhaps in the ester family, electronic factors in the $-CO_2-$ portion of the RCO_2R molecule are more important than shape. The mercaptan family is another case where electronic properties of the functional group, -SH, appear to dominate over shape. The common members of the family smell like a skunk. The skunk actually uses a mixture of mercaptan molecules to defend itself with n-butyl mercaptan, C_4H_9SH, being the major component.

Name _____

Date _____ Lab Section _____

Pre-Laboratory Questions | 22

1. Write an equation that shows the ionization of valeric acid in water. Write equations showing the reaction of valeric acid with sodium hydroxide and with sodium carbonate. Then write the same equations for any carboxylic acid, represented by the general formula RCOOH.

2. Write an equation to illustrate the reactions of 1 mole of malonic acid first with 1 mole and then with 2 moles of sodium hydroxide.

3. Draw the structure of the sodium carboxylate salt of butyric acid, and then write an equation for the reaction that occurs between this salt and dilute aqueous hydrochloric acid.

4. Draw the structure of *n*-butyl benzoate and label the "acid" and "alcohol" portions of this ester. Give the names and structures of the products that would be formed upon hydrolysis of this ester.

5. Use *n*-octyl acetate, a component of orange flavor, to illustrate saponification and hydrolysis reactions. Write the structure of all of the products.

EXPERIMENT

1. Synthesis and Identification of an Ester: Esterification

To 1 mL of concentrated acetic acid and 1 mL of isopentyl (isoamyl) alcohol, $(CH_3)_2CHCH_2CH_2OH$, in a test tube, cautiously add 3 drops of concentrated H_2SO_4. Stir the solution thoroughly, and place the test tube in a boiling water bath for about 10 minutes. Note that two layers form—the top one is ester. Remove a few drops from the top layer with a medicine dropper and place them on a watch glass. See if you can identify the fruit that is associated with the odor. Record your observations and answer the questions. (Answer Question 1, a–f on the Observations and Results sheet.)

Caution: Concentrated H_2SO_4 is dangerous! If you come in contact with the acid, wash it off immediately.

2. Hydrolysis of an Ester: Formation of an Acid and an Alcohol

Add 3 drops of concentrated H_2SO_4 to a mixture of 3 drops of ethyl butyrate, $CH_3(CH_2)_2CO_2CH_2CH_3$ (note the odor), in 3 mL of water. Stir thoroughly. Place the test tube in the boiling water bath for 15 minutes and stir the solution frequently. At the end of this time, use your stirring rod to withdraw a drop of liquid from the test tube. *Allow the stirring rod to cool for a few seconds*, and note the odor. Write an equation for the hydrolysis. Determine which product is responsible for the odor by smelling both authentic products, which are on the reagent shelf. (Answer Question 2, a–f.)

3. Saponification of an Ester

Place 5 drops of methyl salicylate, 3 mL of water, and 1 mL of 20% NaOH (an excess) in a test tube, and heat the test tube in the boiling water bath until the ester layer disappears (about half an hour). Has the odor of the ester disappeared or significantly diminished? Cool the test tube by placing it under a stream of cold water. Then stir in 2 mL of 10% HCl. Check the acidity with litmus and, if necessary, add 10% HCl dropwise until the solution is acidic. Does a precipitate form when the reaction solution is made acidic? What is the name of the precipitate? Record your observations. (Answer Question 3, a–g.)

4. A Carboxylic Acid and Its Salt

A. Solubility of Succinic Acid

Add 1.0 g of succinic acid to 1.5 mL of water in a test tube and stir the solution. Does the acid dissolve? Heat the solution over a burner and stir. Now does the acid dissolve? Cool the test tube in cold water and observe what happens. *Save the mixture*—you will use it for the next part of the experiment. (Answer Question 4A, 1–3.)

B.   Reactions of the Acid and Its Salt

Add with stirring 20% NaOH solution into the cooled succinic acid mixture until the solution is basic to litmus. About 3.5 mL of 20% NaOH will be required.

Neutralize the basic solution by slowly stirring in 4 mL of 6 M HCl. Record and explain your observations. (Answer Question 4B, 1–3.)

> **Dispose of all waste from parts 1-4 in the appropriate containers.**

QUESTIONS AND PROBLEMS

1.   Isopentyl valerate is associated with the smell of ripe apples. Write the structure for this ester, and write an equation for its saponification.

2.   Lactic acid forms when milk is left out in a warm room for several hours. Based on what you have learned in this experiment, give two pieces of evidence indicating that lactic acid is present in "spoiled" milk. Write the structure for lactic acid.

3.   On the basis of the odors of various compounds that you observed in this experiment, what types of organic compounds would you expect to be present in bleu cheese and Limburger cheese?

4.   Calcium propionate is present in bread as a preservative. Write the structure for this salt. Write equations to show how you could make calcium propionate. (Hint: Recall how you made sodium succinate.)

5.   Write equations for the step-by-step ionization of succinic acid in water.

6.   When you saponified methyl salicylate, an acid was produced. Discuss how you would isolate the acid from the reaction mixture and how you could confirm its identity.

7.   Caproic acid, also called hexanoic acid, has an obnoxious odor. Describe how you could remove the odor from clothing on which caproic acid has been spilled. (The acid is not soluble in water.)

EXPERIMENT

Carboxylic Acids and Esters | 22

OBSERVATIONS AND RESULTS

1. Synthesis and Identification of an Ester: Esterification

   a. Give the name and structure of the ester that you synthesized.

   b. What fruit did you associate with the odor of the ester?

   c. Look up the ester and its fruit source in your textbook. (If it is not in your textbook, ask your instructor.) Did you associate the ester with the right fruit? If not, with what fruit is the ester really associated?

   d. Write an equation for this esterification reaction.

   e. Compare the odor of each of the reactants with your ester with respect to *pleasantness* and *intensity*.

   f. Why is the presence of H_2SO_4 necessary for esterification to occur?

2. Hydrolysis of an Ester: Formation of an Acid and an Alcohol

a. What fruit did you associate with the odor of the starting ester?

b. Find the fruit source of this ester in your textbook or ask your instructor. Were you correct in your assignment? If not, with what fruit is the ester associated?

c. What are the names and structures of the products that formed on hydrolysis of ethyl butyrate?

d. Which one of the products is responsible for the odor of the liquid on the stirring rod? What is your evidence?

e. Describe the odor of the liquid on the stirring rod.

f. Give a source of this odoriferous liquid in everyday life.

3. Saponification of an Ester

a. Give the names and uses of two commercial products that contain the ester methyl salicylate.

b. How long did it take for the ester layer to disappear (i.e., for the saponification to occur)?

c. Did the odor of the ester also disappear or diminish? Which?

d. Write an equation for the saponification reaction.

e. Did a precipitate form when the reaction (saponification) solution was made acidic? If so, write the structure of the compound that precipitated.

f. Write an equation for the reaction that occurred when HCl was added to the saponification product.

g. Why was the carboxylic acid that resulted from the saponification process soluble in basic solution but not in acidic solution?

4. A Carboxylic Acid and Its Salt

A. *Solubility of Succinic Acid*

1. Draw the complete structure for succinic acid, showing all of the bonds with dashes.

2. Why is succinic acid called a *di*carboxylic acid?

3. Does succinic acid dissolve in cold water? in hot water? What happened when the test tube was cooled?

B. *Reactions of the Acid and Its Salt*

1. Did succinic acid dissolve when NaOH was added? Why?

2. Record and explain what happened when the basic solution was made acidic with HCl.

3. Write equations for any reactions that occurred in this part of the experiment.

Amines and Amides | 23

OBJECTIVES

1. To learn about the physical and chemical properties of some of the common members of the amine and amide families.

2. To recognize the commercial and physiological value of some amines and amides.

3. To examine those reactions of amines and amides that are important in physiology and biochemistry, particularly in the hydrolysis of proteins.

DISCUSSION

Amines may be viewed as derivatives of ammonia that are formed by successive replacement of the hydrogens of ammonia with alkyl or aromatic groups.

NH_3	CH_3NH_2	$(CH_3)_2NH$	$(CH_3)_3N$
Ammonia	Methylamine	Dimethylamine	Trimethylamine
	(a primary amine)	(a secondary amine)	(a tertiary amine)

As you will observe in today's laboratory, the similarity between amines and ammonia extends beyond the close resemblance of their molecular structures. Some of their physical properties, particularly odor, and many of their chemical reactions are also similar. The three amines shown here are components of herring brine, a fact that may tell you something about their odors. The number of organic groups attached to the nitrogen determines whether an amine is classified as primary (one group), secondary (two groups), or tertiary (three groups). Diamines contain two amino groups in the same molecule. Other important classes of amines are aromatic amines and heterocyclic amines. Some members are shown.

Aniline	β-Naphthylamine	Pyridine
(aromatic-synthesis of dyes)	(aromatic-a potent carcinogen)	(heterocyclic-ring is present in niacin)

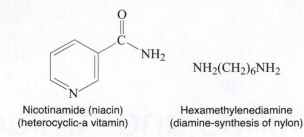

Nicotinamide (niacin)
(heterocyclic-a vitamin)

$NH_2(CH_2)_6NH_2$

Hexamethylenediamine
(diamine-synthesis of nylon)

Probably the two amides encountered most frequently in the beginning laboratory are acetamide and benzamide. An amide of special interest when we are outdoors is *N,N*-diethy-*m*-toluamide, the active agent in the insect repellent Off. Nylon and proteins are prominent members of the amide family. You will study them in later experiments. The functional group that characterizes amides is shown along with some common amides.

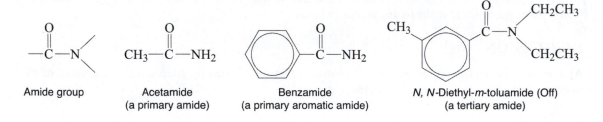

Amide group

Acetamide
(a primary amide)

Benzamide
(a primary aromatic amide)

N, N-Diethyl-*m*-toluamide (Off)
(a tertiary amide)

As with amines, the number of alkyl or aromatic groups on the nitrogen determines whether an amide is primary (no groups), secondary (one group), or tertiary (two groups). The amide urea is unique in having two NH_2—groups on the carbonyl. Urea is a product of the metabolism of amino acids in the cells and is excreted as a waste product in the urine.

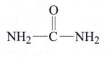

Urea

Reactions of Amines with Water

All three types of amines react with water, as does ammonia, to form the ammonium ion (or substituted ammonium ions) and the hydroxide ion. Equations for the reactions of ammonia and of a primary amine with water are shown here (R– represents an organic group). The solutions are only weakly basic since the equilibria lie far to the left. The basicities of aliphatic amines are comparable to, though in general somewhat greater than, that of ammonia.

$$NH_3 + H_2O \rightleftharpoons NH_4^+ + HO^-$$
Ammonium ion

$$RNH_2 + H_2O \rightleftharpoons RNH_3^+ + HO^-$$
Substituted
ammonium ion

Responsible for
basic litmus test

Reaction of Amines with Acids

Since amines are bases it is not surprising that they react with acids to give salts. Amines, which are usually insoluble in water, generally dissolve in acids through formation of their water-soluble salts.

$$HCl \ + \ RNH_2 \ \longrightarrow \ RNH_3^+Cl^-$$

(Insoluble in water)　　　An aminehydrochloride (water soluble)

When ammonium and substituted ammonium carboxylate salts (except for tertiary amine salts) are heated, water is eliminated and amides are produced. In the following equations, R′ is a hydrogen or an organic group.

$$R'-\overset{\overset{\displaystyle O}{\|}}{C}-OH \ + \ NH_3 \ \longrightarrow \ R'-\overset{\overset{\displaystyle O}{\|}}{C}-O^-NH_4^+ \ \overset{\Delta}{\rightleftharpoons} \ R'-\overset{\overset{\displaystyle O}{\|}}{C}-O^-NH_2 \ + \ H_2O$$

An ammonium carboxylate salt　　　A primary amide

$$R'-\overset{\overset{\displaystyle O}{\|}}{C}-OH \ + \ RNH_2 \ \longrightarrow \ R'-\overset{\overset{\displaystyle O}{\|}}{C}-O^-RNH_3^+ \ \overset{\Delta}{\rightleftharpoons} \ R'-\overset{\overset{\displaystyle O}{\|}}{C}-NHR \ + \ H_2O$$

A substituted ammonium carboxylate salt　　　A secondary amide

Hydrolysis of Amides

Acidic Hydrolysis

Carboxylic acids and ammonium salts form when primary amides are heated in acidic solution. (X represents the anion of a strong acid.)

$$R'-\overset{\overset{\displaystyle O}{\|}}{C}-NH_2 \ + \ HX \ + \ H_2O \ \xrightarrow{\Delta} \ R'-\overset{\overset{\displaystyle O}{\|}}{C}-OH \ + \ NH_4X$$

If the carboxylic acids are sufficiently volatile, they can be detected by their odors. Liberation of ammonia takes place when the hydrolysis solution is made basic.

$$NH_4X \ + \ NaOH \ \longrightarrow \ NH_3 \ + \ NaX$$

Acidic hydrolysis of secondary amides produces carboxylic acids and substituted ammonium salts. Amines are liberated from the substituted ammonium salts in basic solution.

Basic Hydrolysis

Basic hydrolysis of primary amides produces carboxylate salts and ammonia.

$$R'—\overset{\overset{\displaystyle O}{\|}}{C}—NH_2 \ + \ NaOH \ \xrightarrow{\Delta} \ R'—\overset{\overset{\displaystyle O}{\|}}{C}—O^-Na^+ \ + \ NH_3$$

A carboxylate salt

Ammonia can be detected by its odor as the basic hydrolysis proceeds. The carboxylate anions are converted to the only slightly dissociated carboxylic acids by addition of dilute acid.

$$R'—\overset{\overset{\displaystyle O}{\|}}{C}—O^-Na^+ \ + \ HX \ \longrightarrow \ R'—\overset{\overset{\displaystyle O}{\|}}{C}—OH \ + \ NaX$$

Basic hydrolysis of secondary amides gives carboxylate salts and amines. The hydrolysis of peptide bonds (secondary amide bonds) in proteins will be of particular importance in experiment 30. Urea, like other amides, can be hydrolyzed in either acidic or basic solutions. The only unusual aspect of the acidic hydrolysis is the formation of the unstable acid carbonic acid, H_2CO_3, which decomposes immediately into carbon dioxide and water. The sodium salt of carbonic acid, sodium carbonate, is formed in basic hydrolysis.

$$NH_2—\overset{\overset{\displaystyle O}{\|}}{C}—NH_2 \ + \ 2\ NaOH \ \longrightarrow \ Na^+O^-—\overset{\overset{\displaystyle O}{\|}}{C}—O^-Na^+(Na_2CO_3) \ + \ 2\ NH_3$$

ENVIRONMENT, CULTURE, AND CHEMISTRY
Amines and Amides: Two Families with Some Destructive Members

Amines and amides are important members of both our "biological environment" and our "cultural environment." Consider, for example, a few of the members that impact our culture as drugs, both pharmaceutically and recreationally, some with obnoxious odors, and even some with historical significance. Ephedrine, a secondary amine, found in the plant family Ephedraceae and the active ingredient in some nasal decongestants, can be converted to methamphetamine with a set of simple chemical reactions. This devastating recreational drug goes by the name of meth or crystal meth. Recently a major pharmacy chain was indicted and given a huge fine for selling a well-known decongestant, which contained ephedrine, to make meth. The barbiturates (amides), some of the earliest tranquilizers, have been severely abused to avoid the pain of depression and other problems. They were made "famous" as the probable cause of death of Marilyn Monroe. Coniine, a secondary amine found in the common hemlock tree, made history as the poison used to execute the Greek philosopher Socrates. Morphine, a tertiary amine and an important and essential painkiller, is easily converted chemically to heroin; morphine has become the major cash crop of Afghanistan because of the worldwide popularity of heroin. There are many other members of these drug-related families, both good and bad: four common ones are hexamethylene diamine, the monomer in nylon (see Experiment 27); dentists' painkillers benzocaine and xylocaine (amides); the notorious LSD (two amine groups and an amide). Finally, to round out the families there are putrescine and cadaverine, two diamines responsible for the odor of decaying bodies; it is obvious what inspired the names of these two amines.

Name _____

Date _____ Lab Section _____

Pre-Laboratory Questions | 23

1. Name the amides

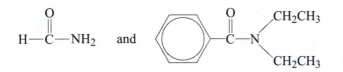

2. Is niacin a primary, secondary, or tertiary amide?

3. Name the products that are formed upon basic hydrolysis of an amide. Name the products resulting from acidic hydrolysis.

4. Write a complete equation for the formation of benzamide from benzoic acid and ammonia.

5. Oxalic acid, HOOCCOOH, is a dicarboxylic acid found in rhubarb and oxalis. Draw a structural formula for the diamide of this acid. Name the diamide.

6. Show the structure of the salt that forms when triethylamine reacts with acetic acid. Write an equation for the reaction that occurs between this salt and sodium hydroxide.

EXPERIMENT

1. Some Common Amines and Amides

You will find several amines and amides on the reagent shelf. Compare the odors of ethylamine and ammonia. Now compare the odors of ethylamine and triethylamine. Note the odor of acetamide. Does it smell like ammonia? Waft all vapors. Test the solubility of *n*-propylamine (if unavailable, try *n*-butylamine) in water by stirring 6 drops of amine into 1 mL of water. Examine the insect repellent Off, noting the odor of the amide *N,N*-diethyl-*m*-toluamide. Record your observations and answer the questions. (Answer Question 1, a–g on the Observations and Results sheet.)

2. The Basicities of Ammonia, Amines, and Amides

A. A Comparison of the Basicities

Place 1 mL of a 1-*M* solution of each of the following in test tubes: ammonia, ethylamine, and acetamide. Determine the pH of the solutions by the following procedure. With your stirring rod transfer 1 drop of the solution to a piece of wide-range pH paper and match the colors immediately. Then use narrow-range pH paper to determine the exact pH. (Answer Question 2A, 1–3.)

B. Reactions of Amines with Acids: Salt Formation

Obtain a drop of ethylamine on a dry, clean stirring rod and hold the amine close to the open mouth of a bottle of concentrated hydrochloric acid, HCl. Watch what happens.[1]

   Stir an amount of benzoic acid the size of a large pea into 2 mL of distilled water. Does the benzoic acid dissolve? Next stir in drops of ethylamine, until the solution is strongly basic to litmus. Note any change in the solubility of benzoic acid. (Answer Question 2B, 1–4.)

> Dispose of all waste from parts 1 and 2 in the appropriate containers.

3. Hydrolysis of Amides

A. Acidic Hydrolysis

Dissolve about 0.5 g of acetamide in 2 mL of 10% sulfuric acid in a test tube and boil the solution over a burner. Identify the odor of the compound that is formed and use moist blue litmus paper to confirm its presence in the vapor. Record your observations. Now add 10% sodium hydroxide solution by drops to the cooled hydrolysis solution until it is basic (about 2 mL). What gas do you smell? Waft the vapors. (Answer Question 3A, 1–2.)

---

[1]Your instructor may wish to demonstrate this reaction.

B. Basic Hydrolysis

Dissolve 0.5 g of urea in 2 mL of 10% sodium hydroxide in a test tube. Heat the solution gently in a flame for a few minutes, and note the odor of the escaping gas. Now make the solution acidic to litmus by adding drops of 10% sulfuric acid solution. Is a gas formed? Waft the vapors and identify the gas. (Answer Question 3B, 1–2.)

QUESTIONS AND PROBLEMS

1. The amide whose structure is shown is added to gasoline as a detergent.

$$CH_3(CH_2)_7CH = CH(CH_2)_7 - \overset{\overset{\displaystyle O}{\|}}{C} - NH(CH_2)_2NH(CH_2)_2OH$$

Name all the functional groups that are present in the molecule. Draw a circle around the amide functional group. Would you expect the detergent to be soluble in water? Explain. Write the structure of the carboxylic acid that would be formed upon acidic hydrolysis of the amide.

2. Suppose the amide in mosquito repellent hydrolyzes on your skin. Write the structures of the products of the hydrolysis. Would you expect these products to be an effective mosquito repellent? Explain.

3. Write a reaction for the hydrolysis of niacin to a carboxylic acid and ammonia.

EXPERIMENT

Amines and Amides | 23

OBSERVATIONS AND RESULTS

1. Some Common Amines and Amides

a. Compare the odors of ethylamine and ammonia.

b. Draw the structures of ethylamine and triethylamine. Identify each as primary, secondary, or tertiary. Compare their odors.

c. Draw the structure of an amine that is an isomer of ethylamine.

d. Does the odor of acetamide resemble ammonia?

e. Is *n*-propylamine soluble in water?

f. Do the ingredients on the label of a container of Off indicate that *N,N*-diethyl-*m*-toluamide is the only component of this insect repellent? If not, what else is present?

g. Describe the odor of *N,N*-diethyl-*m*-toluamide.

2. The Basicities of Ammonia, Amines, and Amides

A. *A Comparison of the Basicities*

1. Record the pH of a 1 *M* aqueous solution of each compound.

$$NH_3\text{_____}CH_3CH_2NH_2\text{_____}CH_3\overset{\overset{\displaystyle O}{\|}}{C}\text{—}NH_2\text{_____}$$

2. Write the names of the compounds above in the order of *increasing* basicity.

3. Which of the organic compounds gives a basic solution when dissolved in water? Write an equation for the reaction.

B. *Reactions of Amines with Acids: Salt Formation*

1. What happens when ethylamine is held close to the mouth of a bottle of concentrated hydrochloric acid?

2. What is the finely divided solid, which looks like smoke? What kind of compound is it? Draw the structure of this solid.

3. Does benzoic acid dissolve in pure water? Does it dissolve when ethylamine is added? Draw the structure of the product of the reaction.

4. Suppose that the solution containing the products of the reaction of ethylamine and benzoic acid was evaporated to dryness and then heated to form an amide. What would be the structure of the amide?

3. Hydrolysis of Amides

A. *Acidic Hydrolysis*

1. Give the name and structure of the volatile compound that you smelled when acetamide was hydrolyzed in acidic solution and explain how blue litmus confirmed its presence.

2. What is the name of the gas you smelled when the hydrolysis solution was made basic? Write an equation for this reaction.

B. *Basic Hydrolysis*

1. What gas (give the name) did you smell when urea was hydrolyzed in basic solution? Write an equation for the reaction.

2. Was a gas formed when the solution was made acidic? If so, write an equation for the formation of this gas.

Vitamin C in Natural and Synthetic Fruit Juices | 24

OBJECTIVES

1. To master a quantitative procedure for analyzing a natural product.
2. To compare the vitamin C content of some familiar fruit juices and a synthetic juice.

DISCUSSION

Vitamin C (ascorbic acid) has long been known for its value in preventing scurvy, a deterioration of the bones and gums. Interest in the many claims of beneficial physiological effects for vitamin C was heightened in 1970 when Nobel laureate Linus Pauling published his book *Vitamin C and the Common Cold.*[1] Pauling suggested that regular intake of the vitamin could be helpful in preventing the common cold and that large doses could treat the symptoms of colds. The prestigious Pauling, a man who did not hesitate to venture beyond the borders of his academic discipline, was not to be taken lightly. Additional investigations into this subject were immediately undertaken.

One of the most comprehensive studies relating to Pauling's claims involved 3615 residents of Toronto, Canada. The subjects were given 1 g of vitamin C daily, the dosage being increased to 4 g per day at the onset of a cold. Some of the subjects were given placebos (tablets without active ingredients but otherwise indistinguishable from the real thing) for control purposes. The study concluded that there was no significant decrease in the number of colds among the group taking the vitamin, but the severity, in time lost from work because of the illness, was reduced by some 30%. There is still much disagreement concerning the effects of vitamin C on colds.

Other claims for beneficial effects of high daily doses of vitamin C are equally interesting. Several investigators report lowered blood cholesterol levels and a decrease in arterial calcification deposits, thus possibly helping to prevent strokes and heart attacks caused by arteriosclerosis. Some report a decreased incidence of wounds' reopening after surgery. Still others point to possible elimination from the body of undesirable metal ions, some fat-soluble carcinogens, and pesticide residues. The subject is still highly controversial and the results of investigations are inconclusive, but the evidence to date is sufficient to warrant further research.

One focal point in the controversy is the proper daily dosage of vitamin C. There appears to be general agreement that the body excretes anything in excess of a 100- or 200-mg daily dose, a quantity called the saturation level. The generally accepted MDR (minimum daily requirement) is 60 mg. However, some experts advocate dosages up to the saturation level. The controversy centers around the massive dosages of 2 to 5 g. The proponents argue that a temporarily high concentration aids in eliminating

---

[1]Pauling, Linus, *Vitamin C and the Common Cold.* San Francisco: W. H. Freeman Publishing Co., 1970.

undesirable ions and molecules. The opponents cannot see how anything beyond the saturation level can be helpful. Considerable time may pass before a clear understanding is reached.

Some investigators consider it preferable to state that deficiencies of important nutrients such as vitamin C may permit or cause physiological dysfunction, rather than to say they are curative when added to the diet. Perhaps in all of the claims for vitamin C, the beneficial action might be ascribed to its role in enabling the body's normal disease-resisting capabilities to function optimally. Possibly vitamin C will prove to be a more important nutrient than has been generally believed.

Vitamin C is found in many fresh fruits and in vegetables such as cabbage, lettuce, Brussels sprouts, broccoli, green beans, and green peppers. It is one of the least stable of the vitamins and may be destroyed in the process used to preserve fruits and vegetables. Because oranges and other citrus fruits are fairly easily stored and shipped, they are a favorite source of vitamin C. British seamen were among the first to recognize the beneficial effects of fresh fruits in preventing scurvy. The practice of taking barrels of limes on long voyages was so common that British sailors became known as "limeys." A commercial source of the vitamin, packaged in tablet form, is rosehips.

Structure of Vitamin C—Reaction with a Red Dye

Vitamin C is water soluble and fairly stable in acid solution, but it is oxidized rather easily in neutral or basic solution, even by oxygen in the air.

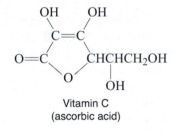

Vitamin C
(ascorbic acid)

Since vitamin C is a reducing agent (i.e., it reacts with oxidizing agents), it can be analyzed readily in the laboratory by its oxidation with iodine, I_2, or with the red dye 2,6-dichlorophenol-indophenol. In the reaction the red dye is reduced to its colorless (leuco) form, and in the process ascorbic acid is oxidized to dehydroascorbic acid.

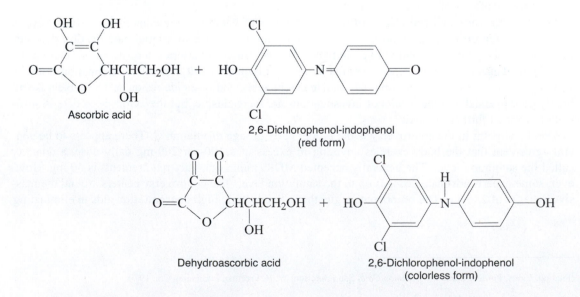

Ascorbic acid

2,6-Dichlorophenol-indophenol
(red form)

Dehydroascorbic acid

2,6-Dichlorophenol-indophenol
(colorless form)

Analysis for Vitamin C

In your experiment today, you will determine the vitamin C content of two natural fruit juices (orange juice and grapefruit juice) and a synthetic juice by titrating them with a solution of red dye of known concentration. The juice will be added to the red dye solution with a buret. When sufficient vitamin C, in the juice, has been added to reduce a fixed amount of red dye, the solution turns colorless.

The vitamin C content of the juices (reported in milligrams per milliliter—mg/mL) can be calculated from the following equation where the equivalency factor equals 0.30 mg of ascorbic acid per milliliter of red dye indicator:

$$\frac{\text{Equivalency factor (mg/mL)} \times \text{volume of red dye (mL)} \times \dfrac{\text{volume of diluted juice in the volumetric flask (mL)}}{\text{volume of original juice in the pipet (mL)}}}{\text{volume of diluted juice added from buret (mL)}}$$

$$= \text{concentration of vitamin C (mg/mL)}$$

ENVIRONMENT, CULTURE, AND CHEMISTRY
Vitamins, Advertising Influence, and Changes in Medical Care

The vitamin and dietary supplement industry has reached the level of $20 billion per year with half of all adult Americans taking a multivitamin per day. Questions continually arise as to whether this consumption of vitamins is necessary to maintain optimal health or is simply a product of intense advertising and promotion for financial gain. Also, take into consideration that medical professionals are often accused of overreliance on pharmaceuticals rather than promoting dietary and lifestyle changes. Recently a group of primary care physicians were asked their opinion on this issue. Most of them did not recommend daily vitamins for healthy patients who eat a well-balanced diet. An adequate diet, according to the professionals, includes fresh vegetables, fruits, whole grains, and meat or a meat substitute. Furthermore, the physicians point out that no daily supplements can substitute for a proper diet since a supplement cannot supply all of the body's needed nutrients in the amounts required. Therefore, since daily vitamins and supplements are not recommended for most people, the conclusion must be drawn that the layperson is making a medical decision. A layperson's decision in the area of medicine can be hazardous. In the case of vitamins and supplements, the possibility of overdose and damage to internal organs exists. One death and eight serious adverse reactions per 40,000 vitamin overdoses have been reported, along with many less serious complications. Another consequence of excess use of vitamins is a waste of money and the necessity of removal at the water treatment plant since some excess vitamins are not stored in the body but excreted in the urine when the Minimum Daily Requirement (MDR) is exceeded.

The post-modern era has a well-recognized suspicion of authority, which includes science and medicine. Hence, the layperson's increased reliance on non-conventional medicine, called holistic or alternative, a form of medicine that often employs natural supplements, herbs, acupuncture, and other procedures has greatly increased in recent years. Some forms of holistic medicine are helpful, and can complement the conventional procedures. However, there is a danger in not relying on the advice of conventional medical doctors (the M.D. degree normally requires seven years of medical education and training beyond the four year undergraduate degree). This dangerous trend was highlighted in the Los Angeles Times. Henry Miller, physician and fellow at Stanford University's Hoover Institute, writes that 40% of American parents have delayed or refused one or more vaccines for their young children; an astonishing 40% of women who may become pregnant take medicine that is contraindicated by the FDA due to the possibility of birth defects in their babies, yet they practice no birth control; women who have a high risk of breast cancer can reduce their risk substantially by taking one of two possible drugs, but one fifth or less take the drugs; hypertension or high blood pressure is a symptom free illness, yet only 10% of patients continue on the drugs for more than one year; asthmatics, diabetics and AIDS patients arbitrarily reduce the dosage or frequency of their therapy or skip it altogether. Clearly, these data indicate that the state of modern, conventional medicine must be seriously assessed.

Name _____

Date _____ Lab Section _____

Pre-Laboratory Questions | 24

1. Name and describe the symptoms of the disease that develops if insufficient vitamin C (ascorbic acid) is present in the diet.

2. What is the name of the indicator in this experiment? Is the indicator also one of the reactants? Were the indicators that you used in previous experiments (phenolphthalein, starch, etc.) also reactants?

3. What functional groups are present in ascorbic acid? In dehydroascorbic acid?

4. The manufacturer of an unsweetened orange juice reports that there are 100 mg of vitamin C in each cup of the juice. How many milligrams of vitamin C are present in each milliliter?

5. Answer the following questions, which relate to experimental procedure.

a. Why is the dye indicator not added from the buret to the juice in the Erlenmeyer flask?

b. Why are the volumes of diluted juices not obtained by using a graduated cylinder?

c. Why is cranberry juice not included in the juices to be titrated in this experiment?

EXPERIMENT

1. Analysis of Vitamin C in Orange Juice

You will need 15 to 20 mL of orange juice.[2] Dilute the juice twenty-fold by transferring 5 mL of it, with a clean pipet, to a 100-mL volumetric flask. Add approximately 1 g of oxalic acid, $H_2C_2O_4 \cdot 2 H_2O$,[3] and dilute with distilled water to the calibration mark. Mix thoroughly by shaking the flask. Rinse and fill a clean 50-mL buret with the diluted juice solution. (Refer to Experiment 14 for detailed instructions on filling and reading a buret and using a pipet.)

To a clean 250-mL Erlenmeyer flask, add exactly 2.00 mL of the standard red dye solution with a pipet. Then add 30 to 40 mL of distilled water so that you have a large enough volume to work with.

Proceed with the titration by adding the orange juice solution, a little at a time, to the red dye (the oxidizing agent) in the flask (figure 24.1). Swirl the flask while you are adding the juice. Place a piece of white paper under the flask so that the end point of the titration can be more easily seen. The end point is signaled by the disappearance of the red color when a single drop of the juice is added. At this point the

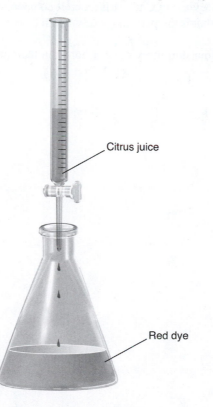

Citrus juice

Red dye

▲ **FIGURE 24.1** Titration of the red dye 2,6-dichlorophenol-indophenol with citrus juice.

[2] We recommend canned juice since it does not contain pulp, which will plug the buret. If fresh juice is used, it must be filtered.

[3] Oxalic acid is added to retard the reaction of vitamin C with oxygen in the air.

amount of vitamin C in the flask is just enough to react with the red dye. Repeat the titration with another sample of red dye. Using the volume (in milliliters) of juice used in your titration and the dilution factor, calculate the concentration of vitamin C (in milligrams per milliliter) in your original orange juice. Average the two results. If they are in poor agreement, do a third titration. (Perform calculations and provide data for 1, a–g on the Observations and Results sheet.)

2. Analysis of Vitamin C in Grapefruit Juice and in a Synthetic Juice

Repeat the titration with grapefruit juice and then with a synthetic (powdered) juice such as Tang.[4] Proceed exactly as you did with the orange juice. (Perform calculations and provide data for Questions 2, a–g; 3, a–k.)

> **Dispose of all waste in the appropriate containers.**

QUESTIONS AND PROBLEMS

1. What acid, in addition to ascorbic acid itself, is present in orange juice?

2. How much grapefruit juice would you have to drink per day before vitamin C would be excreted in your urine? Assume that the grapefruit juice is your only source of vitamin C.

3. Some of the material in a vitamin C tablet is an inert compound, such as starch. Suggest a purpose for adding the starch. What is the percentage of vitamin C in a vitamin C tablet?

4. Survey the persons in your dormitory or your home on their attitude toward vitamin C as a deterrent to colds.

---

[4]The synthetic juice should be prepared as recommended by the manufacturers and then diluted before titration, just as was done with orange juice.

Name _____

Date _____ Lab Section _____

Vitamin C in Natural and Synthetic Fruit Juices

EXPERIMENT 24

OBSERVATIONS AND RESULTS

1. Analysis of Vitamin C in Orange Juice

a. Volume of raw orange juice in the pipet sample _____ mL

b. Volume of diluted juice in the volumetric flask _____ mL

	Run 1	*Run 2*

c. Volume of oxidant (red dye) _____ mL _____ mL

d. Volume of diluted juice required in titration _____ mL _____ mL

e. Vitamin C equivalency of dye

f. Concentration of vitamin C in raw orange juice (show calculations below) _____ mg/mL _____ mg/mL

g. Average concentration of vitamin C in raw orange juice _____ mg/mL

2. Analysis of Vitamin C in Grapefruit Juice

a. Volume of raw grapefruit juice in the pipet sample _____ mL

b. Volume of diluted juice in the volumetric flask _____ mL

	Run 1	*Run 2*

c. Volume of oxidant (red dye) _____ mL _____ mL

d. Volume of diluted juice required in titration _____ mL _____ mL

e. Vitamin C equivalency of dye

f. Concentration of vitamin C in raw grapefruit juice (show calculations below) _____ mg/mL _____ mg/mL

g. Average concentration of vitamin C in raw grapefruit juice _____ mg/mL

3. Analysis of Vitamin C in a Synthetic Juice

a. Volume of synthetic juice in the pipet sample   _____ mL

b. Volume of diluted synthetic juice in the volumetric
   flask   _____ mL

	Run 1	*Run 2*
c. Volume of oxidant (red dye)	_____ mL	_____ mL
d. Volume of diluted juice required in titration	_____ mL	_____ mL
e. Vitamin C equivalency of dye		
f. Concentration of vitamin C in undiluted synthetic juice (show calculations below)	_____ mg/mL	_____ mg/mL
g. Average concentration of vitamin C in undiluted synthetic juice	_____ mg/mL	

h. Rank your juices in the order of decreasing vitamin C content.

i. How much of the orange juice would be needed daily to supply you with 100 mg per
   day of vitamin C? How much grapefruit juice? How much synthetic juice?

j. Judging from vitamin C content only, which is the best breakfast beverage? Explain.

k. According to the label, what ingredients, in addition to vitamin C, are present in the
   synthetic juice? Do you think that they are also present in orange juice? Are the other
   ingredients in the synthetic juice beneficial in your diet? Are they detrimental?

Preparation and Examination of a Drug: Aspirin | 25

OBJECTIVES

1. To prepare the common drug aspirin.
2. To examine the chemical and physical properties of aspirin.
3. To compare the "synthesized" aspirin with commercial aspirin.
4. To consider how aspirin exerts its medicinal effect.

DISCUSSION

Aspirin is certainly the most widely used medicinal agent ever discovered. The world production is estimated at 90,000,000 lb per year. Common dosages in the United States are 81 and 325 mg. Aspirin has many uses: an antipyretic (reduces fever); an anti-inflammatory; an analgesic, relieving the mild pain of headaches, neuralgia, and rheumatism; reducing the chance of heart attacks and strokes; perhaps even preventing certain cancers. Over the past few years, aspirin has faced increased competition from ibuprofen and acetaminophen, with each having advantages over the other depending, on the medical situation.

The Relationship of Aspirin to Salicylic Acid

Aspirin is an ester derivative of salicylic acid. Its chemical name is acetylsalicylic acid; its common name is derived from the old German version of the name, *acetylspirasaeure.*

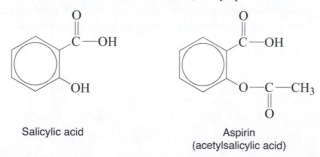

Salicylic acid

Aspirin
(acetylsalicylic acid)

Aspirin was first marketed commercially by the Bayer Company in Germany in 1899. The discovery of its important medicinal value was actually a result of earlier studies on the parent compound, salicylic acid, which was shown to have medicinal value, but also some undesirable side effects.

Side Effects of Aspirin

Actually, aspirin is not without its own side effects. Studies have shown that some hemorrhaging of the stomach wall occurs with even the normal dosage. The average person loses from 0.5 to 2.0 mL of blood when two aspirin tablets are swallowed. This amount of damage is of no consequence, except for persons who use large quantities of aspirin, for example, arthritics. It is known that the damage done by aspirin is caused by the un-ionized carboxylic acid form and not by the ionized carboxylate anion form, since only the former can penetrate the lining (mucosa) of the stomach.

Examination of the equilibrium equation suggests that if the pH of the stomach, which is normally quite low (acidic), is buffered near the neutral point, the equilibrium shifts to the right, resulting in a predominance of the carboxylate anion form of aspirin.

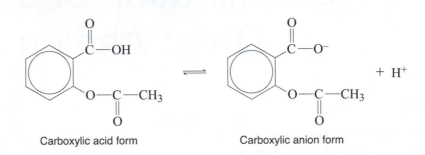

Carboxylic acid form   Carboxylic anion form

There should be less damage of the mucosa under these conditions. The only problem with commercial buffered aspirin is that the quantity of buffer in a tablet or two generally is not sufficient to affect the pH of the stomach fluid significantly.

How Aspirin Exerts Its Effect

At low dosages, aspirin inhibits an enzyme involved in the formation of a molecule called thromboxane, which causes blood platelets to coagulate. In the absence of thromboxane, there are fewer blood clots and, hence, less chance for blockage of the arteries, which can lead to heart attacks or strokes. At higher doses of aspirin, as at lower doses, thromboxane is not formed, and a second enzyme is inhibited, preventing the formation of another group of molecules called the prostaglandins. Prostaglandins are thought to signal the nerves' pain receptors and to be involved in the development of fever.

The Synthesis of Aspirin

In today's experiment you will synthesize aspirin, study some of its properties, and compare it with commercial aspirin. You will prepare acetylsalicylic acid by heating salicylic acid with acetic anhydride in the presence of the catalyst phosphoric acid, H_3PO_4.

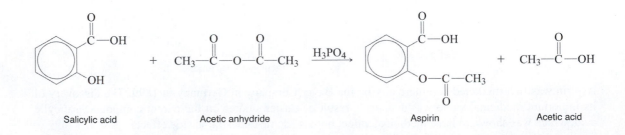

Salicylic acid   Acetic anhydride   Aspirin   Acetic acid

Although you may not be familiar with this type of esterification, it is an excellent method for synthesizing an ester (e.g., aspirin).

The Quality and Quantity of Your Drug

As is true of all pharmaceutical manufacturers, you must be concerned with the quality (purity) and quantity (yield) of your drug. The percentage yield represents the extent, expressed as a percentage, to which the reactants are converted into the isolated product. It is calculated by the equation,

$$\text{Percentage yield} = \frac{\text{actual yield}}{\text{theoretical yield}} \times 100.$$

The actual yield is the weight (or volume, for a gas) of product actually isolated, and the theoretical yield is the weight (or volume, for a gas) of product that would be formed if the reaction went to completion as indicated by the balanced equation. For example, suppose a student isolated 3.5 g of aspirin (actual yield) from a reaction of 4.0 g of salicylic acid with excess acetic anhydride. What is the theoretical yield of aspirin? What is the percentage yield? We see from the balanced equation that 1 mole of salicylic acid (138 g) produces 1 mole of aspirin (180 g).

$$4.0 \text{ g of salicylic acid} + \text{acetic anhydride} \xrightarrow{\text{H}_3\text{PO}_4} ? \text{ g of aspirin} + \text{acetic acid}$$
$$\text{(1 mole = 138 g)} \qquad\qquad\qquad\qquad\qquad \text{(1 mole = 180 g)}$$

Then 4.0 g (4.0/138 mole) of salicylic acid should produce the same fraction (4.0/138) of a mole (180 g) of aspirin. Thus, the theoretical yield would be

$$\frac{4.0 \text{ g}}{138 \text{ g}} \times 180 \text{ g} = 5.2 \text{ g of aspirin}$$

The percentage yield then is

$$\frac{3.5 \text{ g}}{5.2 \text{ g}} \times 100 = 67.3\%$$

Now consider the purity of your aspirin. What is the most likely contaminant? Acetic anhydride is unlikely to be present in the aspirin since it is converted to acetic acid when water is added. Acetic acid is soluble in water and would be washed away in the wash water. However, any salicylic acid that did not react might mix with the aspirin because salicylic acid is quite insoluble in water. You will test for its presence in your aspirin. How? Acetylsalicylic acid is both a carboxylic acid and an ester. You will study reactions of both functional groups.

ENVIRONMENT, CULTURE, AND CHEMISTRY
Medicinal Drugs from Nature

The majority of the medicinal drugs in use today have their origin in nature. In fact there is currently a mad scramble to examine as many fungi, plants and trees as possible before they become extinct to learn whether they might be sources of useful drugs. Current estimates place 12% of all plant species in danger of extinction. The rush to find drugs in nature was the basis of the intriguing movie, *Medicine Man*, showing a scientist (played by Sean Connery) in the jungle, armed with scientific instruments (a great stretch of the imagination), eagerly searching for a particular drug. The actual isolation of a drug and identification of its chemical structure is a relatively recent phenomenon. Chewing the leaves of the willow tree was probably known in prehistoric times to relieve pain and fever. Yet it was only in the late 1800s that salicylic acid was isolated from the leaves and pinpointed as the source of the medicinal effect. Salicylic acid was soon converted chemically to aspirin, much as you will do in lab today; aspirin was determined to have far less damage on the lining of the stomach than salicylic acid. Penicillin was discovered in a mold. In 1928, Alexander Fleming observed that a particular bacterium was destroyed by a mold, *Penicillium notatum*. Approximately ten years later, Howard Florey and Andrew J. Moyer, recognizing the importance of Fleming's discovery, greatly increased the production of the new drug, called penicillin, just in time to save the lives of many World War II troops. Thus, the world of antibiotics was born. The Nobel Prize was awarded to the three scientists for their monumental achievement. Penicillin was eventually isolated in pure form and converted by scientists to new, molecularly altered "penicillins," an ongoing task since bacteria rapidly develop resistance to each new penicillin. More recently, the bark of the Pacific yew tree gave us the anti-cancer drug Taxol. Taxol has become a lifesaver for victims of cancer, particularly ovarian cancer. Taxol has now been synthesized by chemists and the destruction of the yew tree is no longer required. Morphine, an important painkiller made from the poppy plant comes to mind at this time because it is the "cash crop" in war-torn Afghanistan. Positive claims for "drugs from nature," however, cannot necessarily be taken as fact. Laetrile (also marketed as amygdalin and vitamin B17), an extract from apricot pits, is an example of a drug that has long been touted as a cure for cancer. First used in Russia in the 1840s without success, laetrile is still available today, particularly in Mexico, despite a negative decision by the U.S. Supreme Court upholding the findings of the National Cancer Institute, the Mayo Clinic, and prestigious universities over many years, which have clearly established that it does not cure cancer. The death of actor Steve McQueen in 1980, while undergoing treatment for cancer with laetrile in Mexico, certainly underscored its lack of efficacy. The desperate desire for a cure in the face of a terminal illness is understandable: consider the death of beloved Coretta Scott King, widow of Martin Luther King, Jr., in a holistic medicine therapy center in Mexico in 2006, which was later reported to be unlicensed and was closed shortly after her death.

Pre-Laboratory Questions | 25

1. Draw the structures of salicylic acid and acetylsalicylic acid (aspirin) and identify all functional groups. Which of the two compounds is a phenol?

2. Write an equation for the hydrolysis of aspirin as it occurs in the intestine.

3. In this experiment excess acetic anhydride is converted to acetic acid when water is added to the reaction product. Write an equation for this reaction.

4. Discuss why it is important to determine the yield in a synthesis reaction.

EXPERIMENT

1. The Synthesis of Aspirin

Weigh 1.5 g of salicylic acid and place it in a 125-mL Erlenmeyer flask. Then add 4 mL of acetic anhydride and 4 drops of 85% phosphoric acid. *Measure the acetic anhydride under the hood since it is highly irritating to the nose.* Stir the reaction mixture thoroughly. Heat the Erlenmeyer flask in a boiling water bath for 5 minutes, and stir it frequently (figure 25.1). Remove the flask from the bath, and stir 3 mL of water at once into the hot mixture. Continue to stir for a couple of minutes in order to destroy any excess acetic anhydride, and then continue to stir while you add 30 mL of water. At this point, aspirin begins to precipitate from the solution. Complete the precipitation by cooling the flask in an ice-water bath for about 5 minutes (keep stirring).

   Isolate your aspirin by pouring the solution into a Büchner funnel (figure 25.2) *that is operating properly.* The operating instructions are as follows.

   A. The aspirator must pull a strong vacuum. To test it, start the water running and put your hand over the top of the filter flask (with the funnel removed) for 10 to 15 seconds to see if a vacuum develops.

   B. Your filter paper should fit exactly the bottom of the Büchner funnel; it must not extend up the sides of the funnel. You may have to cut the paper to make it fit.

   C. Wet the filter paper in the Büchner funnel with water immediately before pouring in the aspirin solution.

   D. *With the aspirator on full,* pour the aspirin product rapidly into the center of the filter paper. Use a spatula to transfer any crystals remaining in the flask.

▲ **FIGURE 25.1** The preparation of aspirin. Salicylic acid reacts with acetic anhydride at 100°C to produce acetylsalicylic acid.

Wash your aspirin crystals thoroughly with 10 mL of ice (or at least cold) water by pouring the water slowly over the crystals as you stir them. Distribute the crystals across the bottom of the Büchner funnel with a spatula and dry them by pulling air through the Büchner funnel for a few minutes. Stir the crystals frequently with your spatula so that they will all be exposed to the passing air. Further dry the crystals by pressing them between two pieces of folded paper towels.

Weigh your aspirin and determine the yield. (Record your results in Question 1 on the Observations and Results sheet.)

2. Contaminants in Aspirin

Use the iron(III) chloride test, which you studied in the experiment on phenols (Experiment 20), to determine if any salicylic acid is present in your aspirin and in commercial aspirin. Proceed as follows. Add about 0.1 g of salicylic acid (for comparison) to 2 mL of 1% iron(III) chloride solution in a test tube. Then add about 0.1 g of your aspirin to the same amount of iron(III) chloride in another test tube, and finally add about 0.1 g of finely crushed commercial aspirin to 2 mL of iron(III) chloride solution in a third test tube. Stir each solution thoroughly. Note the color, if any, in each case. Is either your aspirin or commercial aspirin contaminated with salicylic acid? (Answer Question 2, a–c.)

Commercial aspirin is only 70% to 90% acetylsalicylic acid; most of the remainder is a binder, usually starch, which holds the tablet together. To confirm that starch is the binder, dissolve some commercial aspirin (the size of a pea) in a couple of milliliters of water, while stirring. Now add 1 drop of dilute iodine solution, and see whether the blue-black starch-iodine complex is formed. (Answer Question 2d.)

3. Aspirin as an Acid

Add 100 mg (an amount the size of a large pea) of your aspirin, commercial aspirin, Bufferin, and salicylic acid to 2 mL of boiled distilled water in four separate test tubes. Stir the solutions. Use narrow-range pH

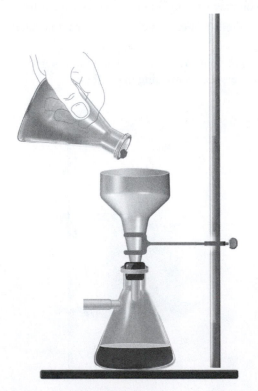

▲ **FIGURE 25.2** Collecting a precipitate in a Büchner funnel.

paper to determine the pH of the solutions, as described in Experiment 23. Add 100 mg (a pea-sized amount) of solid sodium bicarbonate to your aspirin solution. Is a gas formed? Record your observations. (Answer Question 3, a–c.)

4. Aspirin as an Ester: A Test for Aspirin

Dissolve 0.2 g of crushed commercial aspirin in 1 mL of 10% sodium hydroxide in a test tube (test the solution with litmus paper to confirm that it is basic), and heat the test tube in a boiling water bath for several minutes. To the hot solution add concentrated hydrochloric acid, stirring it in drop by drop, until the solution is just acidic to litmus. Smell the contents of the test tube. Do you smell acetic acid? Now add several drops of iron(III) chloride solution. Does a colored complex form? Record your observations. (Answer Question 4, a–f.)

> **Dispose of all waste in the appropriate containers.**

QUESTIONS AND PROBLEMS

1. Why is it important to drink water immediately after taking aspirin? (How is the stomach wall affected?)

2. Sodium salicylate has been used as an analgesic. Draw the structure of this compound and show how you could prepare it from salicylic acid.

3. The analgesic Tylenol is often taken by persons who are allergic to aspirin. Tylenol contains acetaminophen (structure shown) as the active ingredient.

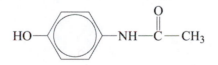

Is the structure of acetaminophen similar to the structure of aspirin? In what way? Would acetaminophen give a positive phenol test? What products would be obtained if acetaminophen were hydrolyzed in acidic aqueous solution?

4. The makers of the analgesic Bufferin claim that their product buffers the fluid of the stomach at a higher pH, thereby avoiding the side effects of aspirin. What buffer is present in Bufferin? Is there enough buffer present in a couple of tablets to significantly raise the pH of the stomach fluid when taking aspirin?

EXPERIMENT

Preparation and Examination of a Drug: Aspirin

25

OBSERVATIONS AND RESULTS

1. The Synthesis of Aspirin

Record the weight of your aspirin, and calculate the percentage yield. Base the yield on salicylic acid since excess acetic anhydride was used. Show all calculations.

Weight of aspirin   _____ g

Yield of aspirin   _____ %

2. Contaminants in Aspirin

a. What was the color of the salicylic acid–iron(III) chloride test solution?

b. On the basis of color, is either your aspirin or commercial aspirin contaminated with salicylic acid? What is your evidence for contamination?

c. Acetic acid was also formed in the synthesis of aspirin and would be present with aspirin in the reaction mixture. Why is it unlikely that your isolated aspirin is contaminated with acetic acid?

d. Is starch the binder in the commercial aspirin in your laboratory? What is your evidence?

3. Aspirin as an Acid

a. Draw the structure of aspirin. Circle and name the functional group that makes aspirin an acid.

b. Record the pH of each of the compounds.

Your aspirin   _____

Commercial aspirin   _____

Bufferin   _____

Salicylic acid   _____

c. Write an equation that shows why aspirin gives an acidic solution when dissolved (or partially dissolved) in water.

d. Did you find that Bufferin is less acidic than aspirin?

   Did your aspirin react with sodium bicarbonate to form a gas? If so, what is the gas?

e. Write an equation for the reaction between aspirin and sodium bicarbonate.

4. Aspirin as an Ester: A Test for Aspirin

a. Draw the structure of aspirin and circle the functional group that makes aspirin an ester.

b. Write an equation for the rapid reaction (call this reaction 1) that occurs when aspirin (an acid) dissolves in sodium hydroxide (a base) solution.

c. Now write an equation for the slower reaction (reaction 2) between aspirin (an ester) and sodium hydroxide in the boiling water bath.

   What is the name of reaction 1?
   What is the name of reaction 2?

d. Did you smell acetic acid after the reaction mixture was neutralized with hydrochloric acid?

e. Can the presence of acetic acid be used as a test for aspirin? Explain.

f. What was the color of the neutralized solution after the addition of iron(III) chloride solution? What organic compound (name and structure) was responsible for this color? What functional group was responsible for this colored complex?

Carbohydrates | 26

OBJECTIVES

1. To become familiar with several of the common carbohydrates and with their important physical properties.
2. To learn the significant differences in the physical and chemical properties of monosaccharides, disaccharides, and polysaccharides.
3. To relate the reactions of carbohydrates to the fundamental chemistry of functional groups studied in previous experiments.
4. To learn some of the reactions of carbohydrates that are important in metabolism.

DISCUSSION

Carbohydrates represent the primary source of energy required by our bodies. Active persons burn large amounts of carbohydrates, but excessive carbohydrates in the diet are converted to fats and stored. Carbohydrates are present in a wide assortment of foods: all kinds of grains, potatoes, lean meat, and even codfish and spinach. Probably you have seen the analysis of the carbohydrate content of your favorite breakfast cereal on the side panel of the box. In this experiment, and in later ones also, you will analyze some familiar food products.

Classification of Carbohydrates

The broad, confusing field of carbohydrates can be narrowed and simplified considerably by grouping into three classes: monosaccharides, disaccharides, and polysaccharides. Some common carbohydrates, their classes, their sources, and the products they yield upon hydrolysis are shown in table 26.1. (How many of them are you familiar with from your everyday experiences?)

Disaccharides contain two monosaccharide units joined together by an acetal (glycoside) bond. Polysaccharides are polymers in which the monomers are monosaccharides. Glucose (in honey), maltose (in corn syrup), and amylose (in potatoes) are common examples of the three classes of carbohydrates. Note the glucose units in maltose and amylose (see the structures).[1]

---

[1]In the structures of the carbohydrates shown in this experiment, a carbon with only three bonds indicated is understood to have a hydrogen also.

TABLE 26.1 Some Common Carbohydrates

Carbohydrate	Class	Source	Monosaccharides Obtained by Hydrolysis
Glucose (dextrose)	Mono-	Honey	—
Fructose (levulose)	Mono-	Honey	—
Galactose	Mono-		—
Mannose	Mono-		—
Xylose	Mono-		—
Maltose	Di-	Corn syrup	Glucose
Sucrose	Di-	Sugarcane, sugar beets, maple sap, and other plant fluids	Fructose, glucose
Lactose	Di-	Milk	Glucose, galactose
Starch (amylose and amylopectin)	Poly-	Seeds, tubers	Glucose
Glycogen	Poly-	Liver tissue, muscle tissue	Glucose
Cellulose	Poly-	Plant cells, wood, cotton, paper	Glucose
Inulin	Poly-	Artichokes, dahlias	Fructose
Xylans	Poly-	Wood, straw, corncobs, grain hulls	Xylose

Physical Properties

All monosaccharides and disaccharides, and some polysaccharides, are soluble in water but insoluble in organic solvents. Carbohydrates, which are actually polyalcohols, form hydrogen bonds with water. Concentrated sugar solutions (for example, honey) are viscous syrups that are supersaturated and crystallize slowly. Sucrose (table sugar) and many other members of the family have a sweet taste. The degree of sweetness depends on the particular structure. Some compounds, with unrelated structures (for example, saccharin) are far sweeter than sugar. You will examine this physical property of several carbohydrates in today's experiment.

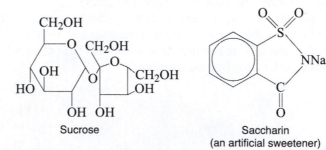

Sucrose

Saccharin
(an artificial sweetener)

Reducing Sugars

All monosaccharides and many disaccharides reduce weak oxidizing agents such as the Cu^{2+} in Fehling's reagent.[2] These carbohydrates are called *reducing sugars*. To function as a reducing sugar, a carbohydrate must have an aldehyde functional group[3] or a hemiacetal functional group, which can open to become an aldehyde. Of the three forms of glucose, only the open-chain (acyclic form) is oxidized by Fehling's reagent.

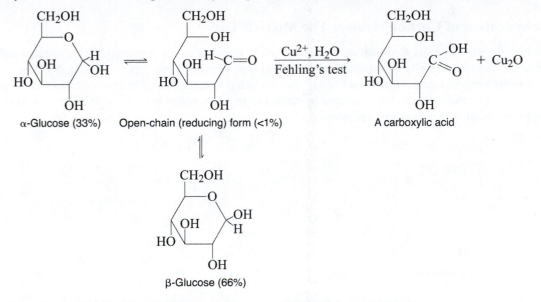

You will test the reducing capability of several carbohydrates in today's experiment.

Hydrolysis of Disaccharides and Polysaccharides

Hydrolysis of the acetal bonds in disaccharides (see maltose) gives monosaccharides. Hydrolysis of polysaccharides, such as amylose and amylopectin, proceeds in stages and yields several products.

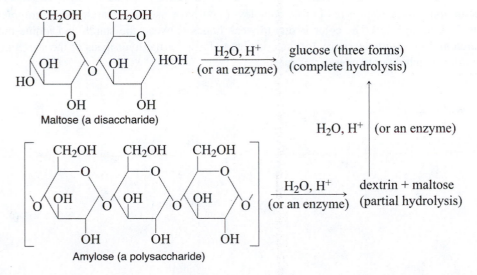

---

[2]As a background for this experiment, review Fehling's (Benedict's) test and the hydrolysis of acetals in experiment 21.

[3]Fructose is a special case in which a ketone is readily oxidized because the hydroxyl groups adjacent to the carbonyl group are activated.

Cleavage of all of the acetal bonds (*complete hydrolysis*) produces only glucose. *Partial hydrolysis* of a few random bonds gives many shortened starch molecules, called dextrins, and some maltose. Maltose molecules result whenever an acetal bond remains between two glucose units. Corn syrup, a mixture obtained from the partial hydrolysis of corn starch, contains considerable maltose (see table 26.1). In this experiment, you will determine the effect of hydrolysis on the reducing capability of carbohydrates. What do you predict the effect will be?

Dehydration of Carbohydrates: The Molisch Test

Carbohydrates, like most alcohols, undergo dehydration reactions in the presence of concentrated sulfuric acid.[4] Pentoses (five-carbon sugars) give furfural (see structure), and ketohexoses and aldohexoses give substituted furfurals. This is the basis of the Molisch test, a general carbohydrate test in which furfural and the substituted furfurals formed by dehydration of monosaccharides[5] with sulfuric acid react with α-naphthol to form colored compounds.

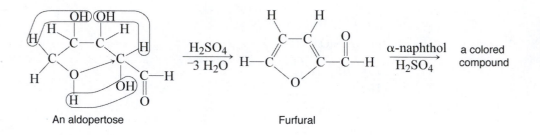

An aldopertose　　　　　　　　　　Furfural

You will use the Molisch test to detect carbohydrates in several common foods.

The Starch-Iodine Complex

Starches consist of a mixture of amylose and amylopectin. Potato starch, for example, is 90% amylopectin and 10% amylose. Amylose reacts with iodine, I_2, to give an intense blue-black complex, and amylopectin and iodine form a red-purple complex. (You have used the starch-iodine complex in experiments 7, 8, 9, and 25.) The color is due to weak bonds between the starch and iodine molecules. Large starch molecules are necessary for the complex to form. Partial hydrolysis of the starch molecules gives dextrins, which do not undergo this reaction.

---

[4]Refer to experiment 20 for a review of the dehydration reaction.

[5]Disaccharides and polysaccharides are hydrolyzed to monosaccharides, which are then dehydrated.

ENVIRONMENT, CULTURE, AND CHEMISTRY
Carbohydrates as a Source of Fuel

Carbohydrates are often discussed as key substances in our attempts to achieve a sustainable energy source, free of dependence on foreign oil. Because the plant/tree world is mostly carbohydrate, there is an abundance of this material. Each year the plants/trees of earth use solar energy to trap carbon dioxide and synthesize some 100 billion tons of carbohydrate material. The microorganisms of the ocean make a similar quantity of carbohydrate each year. Potentially, this is a huge amount of material/energy but living things consume most of it every year, eventually putting much of it to human use. The total production of corn, wheat, and rice (mostly carbohydrate) in 2007 was 2.3 billion tons. But given the fact that the nations of the world consume 3.5 billion tons of non-renewable petroleum each year, and not considering the energy conversion factor between carbohydrates and petroleum, this still leaves a tremendous gap between the energies from surplus carbohydrate material produced and petroleum consumed, providing an enormous challenge to find renewable replacements. With the scarcity of land that is suitable for agriculture to grow crops, it is unlikely that more than a fraction of the petroleum/carbohydrate gap can be met by agriculture. One of the ideas being tried on a large scale is the conversion of corn to ethanol as an alternative fuel. Much energy is consumed to grow and process the corn and so far the gain is unimpressive, but the experience gained from research in this area is important. There is a real possibility of producing larger amounts of biomass (carbohydrate) on marginal land if the appropriate plant/land combination is discovered. Direct energy-capture methods such as solar and wind hold promise as a partial source of some of our energy needs. Even energy from nuclear material must be considered, if it can be done safely. Perhaps with time and with more study, these methods can begin to fill the large gap described above and eventually even eliminate our dependency on petroleum, both foreign and domestic.

Pre-Laboratory Questions | 26

1. Give the names of a common aldohexose and a common ketohexose.

2. Do you predict that maltose will reduce cupric ion in Fehling's reagent in today's experiment? Explain.

3. Can a dilute iodine solution be used to distinguish between amylose and amylopectin? Explain.

4. Compare the structures of dextrin and maltose. Compare the structures of dextrin and amylose.

EXPERIMENT

1. An Important Physical Property—Sweetness

OPTIONAL EXPERIMENT: Your instructor may not want you to taste chemicals in the laboratory.

Determine the relative degree of sweetness of the following carbohydrates by placing a small amount of each on your tongue: glucose, lactose, fructose, sucrose, maltose, and starch. Rinse your mouth with water after each test, using a clean styrofoam cup. Now compare saccharin with the sweetest carbohydrate. (Answer Question 1, a–d on the Observations and Results sheet.)

2. Carbohydrates as Reducing Sugars

Perform Fehling's test on 2% solutions of the following carbohydrates: lactose, sucrose, glucose, fructose, and starch. Also test solutions of honey and corn syrup. Proceed as follows: To each clean test tube add 4 drops of the carbohydrate solution and 2 mL of Fehling's reagent (remember to prepare Fehling's reagent by mixing equal amounts of the A and B solutions). Prepare the honey and corn syrup solutions by estimating 2 drops of each and mixing them with 2 drops of water. Stir thoroughly. Place the test tubes in a boiling water bath for 10 minutes. (Answer Question 2, a–e.)

> **Dispose of waste in the appropriate container.**

3. Hydrolysis of Disaccharides and Polysaccharides

You will study the hydrolysis of starch and sucrose by testing for the formation of reducing sugar with Fehling's reagent. You will use the iodine test to provide further confirmation of the hydrolysis of starch. Refresh your memory of the color of the starch-iodine complex by adding 1 drop of dilute iodine solution to 10 drops of the 2% starch solution on a watch glass. What is the color of the complex? (Answer Question 3a.)

Now perform the hydrolysis in the following manner. Stir 2 drops of concentrated hydrochloric acid into 2 mL of the 2% starch solution and into 1 mL of the 2% sucrose solution in a second test tube. Label the test tubes and place them in a boiling water bath for 15 minutes.

Next, carry out the iodine test on 10 drops of the hydrolyzed starch solution on a watch glass. Has hydrolysis of the starch occurred? (Answer Question 3b.) Test further for hydrolysis of the starch and the sucrose solutions by neutralizing the acid in each solution with 10% sodium hydroxide (add drop by drop, stirring, until the solution is just basic to litmus) and adding 2 mL of Fehling's reagent. Heat in the water bath for 10 minutes. Does Fehling's test indicate that hydrolysis has occurred? (Answer Question 3, c–d.)

4. The Molisch Test: A General Carbohydrate Test

First conduct the Molisch test with glucose to serve as a point of reference. Into 10 drops of a 2% glucose solution in a test tube, stir 2 drops of Molisch reagent (10% α-naphthol in ethanol). Now tilt the test tube and carefully drop 10 drops of concentrated sulfuric acid, H_2SO_4, down the side of the test tube, so that the sulfuric acid forms a layer on the bottom (figure 26.1). Note the color at the interface between the acid and aqueous solutions. Now perform the Molisch test on 10 drops of each of the following solutions: a nonreducing disaccharide (2%), a ketose (2%), honey (50% with water), corn syrup (50% with

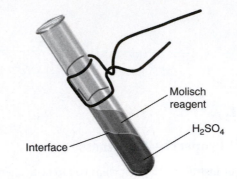

▲ **FIGURE 26.1** The Molisch test. The color at the interface indicates the presence of a carbohydrate.

water), and flour (suspension made by mixing a small amount (size of a small pea) with 10 drops of water). (Answer Question 4, a–c.)

> **Caution:** Concentrated H_2SO_4 is dangerous! If you come in contact with the acid, wash it off immediately.

5. Tests for Starch in Food Products

Place a few drops of the iodine test solution on the following food products on a watch glass: a slice of potato, a piece of white bread, a piece of cracker, some powdered macaroni, and a powdered white bean. (Answer Question 5.)

QUESTIONS AND PROBLEMS

1. List several food products, in addition to the ones that you studied in the laboratory, which would give a positive iodine-starch test for carbohydrate.

2. Suggest how you might make "honey" from table sugar.

3. Draw the structure of a segment of an amylose molecule, and show how water breaks the acetal bonds.

4. Draw the structure of a segment of a molecule of inulin. On the basis of today's experiment, suggest a way to determine whether a compound is amylose or inulin.

5. Compare the product that would be obtained from hydrolysis of a 50:50 mixture of inulin and amylose with the hydrolysis product of sucrose. How would it compare with honey?

6. Explain why carbohydrates would be of great importance in a world with little or no petroleum.

EXPERIMENT

Carbohydrates | 26

OBSERVATIONS AND RESULTS

1. An Important Physical Property—Sweetness

a. Rank the disaccharides in decreasing order of sweetness.

b. Rank all of the carbohydrates examined in decreasing order of sweetness.

c. Is sweetness a reliable measure of the carbohydrate content of a food? Explain.

d. How do the sweetness of saccharin and the sweetness of the sweetest carbohydrate compare?

2. Carbohydrates as Reducing Sugars

a. Classify the sugars tested as reducing or nonreducing.

        Reducing sugars   *Nonreducing sugars*

b. Did honey give a positive Fehling's test? Give the names and structures of the carbohydrate(s) which was (were) responsible for the test.

c. Did corn syrup give a positive Fehling's test? What carbohydrate(s) was (were) responsible for this test?

d. Write equations for the reactions of *two* carbohydrates that gave positive Fehling's tests.

e. Explain why two of the carbohydrates failed to give positive tests. (The reasons for the failure are different for the two carbohydrates.)

3. Hydrolysis of Disaccharides and Polysaccharides

a. What was the color of the initial iodine test on the starch solution?

b. After the starch solution was heated in the water bath for 10 minutes, did the iodine test indicate that hydrolysis had occurred? Did the Fehling test confirm this? Give the evidence for both of your answers.

c. Did Fehling's test indicate that hydrolysis of sucrose had occurred after 10 minutes? What is your evidence?

d. Give the names of the products that were formed upon hydrolysis of sucrose and starch.

   Sucrose:

   Starch:

4. The Molisch Test: A General Carbohydrate Test

a. Give the name of each carbohydrate and the exact color obtained for each in the test.

Type of Carbohydrate	*Name of Carbohydrate*	*Color*
A nonreducing disaccharide	_____	_____
A ketose	_____	_____
Honey	_____	_____
Corn syrup	_____	_____
Flour	_____	_____

b. What type of reaction occurs with each carbohydrate in the Molisch test? What organic product is formed? Is the same organic product formed for each carbohydrate? Explain on the basis of your observations.

c. Draw a segment of the structure of a typical molecule of the carbohydrate in flour.

5. Tests for Starch in Food Products

Which of the food products gave a positive test for starch?

Natural and Synthetic Polymers: The Preparation of Nylon | 27

OBJECTIVES

1. To learn some distinctions between natural and synthetic polymers.
2. To examine the monomers of two common fibers.
3. To become familiar with some of the physical and chemical properties of two common polymers.
4. To prepare the important polymer nylon.

DISCUSSION

Nylon and cellulose are common polymers that make numerous contributions to our everyday lives. Nylon is a synthetic polymer. Cellulose occurs abundantly in nature as the main structural material of most plant tissues. Cotton is nearly pure cellulose. Both nylon and cotton are used in fabrics for clothing. Solid items such as gears, combs, and knobs are molded from nylon. Cellulose is converted into cellulose acetates, rayon, and other important fibers and films. Guncotton, the chief propellant in smokeless powder, is manufactured by treating fibers of cotton with a mixture of nitric and sulfuric acids under appropriate conditions. In today's experiment you will synthesize nylon and study the hydrolysis of cellulose.

What Are Polymers?

Polymers are high-molecular-weight molecules that are composed of simple, repeating units. Small molecules that are joined together by covalent bonds to become the repeating units of polymers are called monomers; hence, the repeating units are called monomeric units. When you scan the molecular structure of a polymer, you can usually recognize the monomeric units. From the structure of the recurring unit, you can deduce the structure of the monomer. Many polymers are derived from only one kind of monomer and, hence, contain only a single recurring unit. Others contain two or more kinds of monomeric units.

The Synthesis of Nylon

Both nylon and cellulose are examples of condensation polymers. A condensation polymer is one in which the repeating units have been formed from the monomers by elimination of a small molecule such as water or an alcohol.[1] For example, the E. I. du Pont de Nemours Company, which pioneered nylon research, synthesizes Nylon 6-10 by *condensing* sebacic acid and hexamethylenediamine to give a polyamide (nylon), with the elimination of water.

$$HO-\overset{\overset{\textstyle O}{\|}}{C}(CH_2)_8\overset{\overset{\textstyle O}{\|}}{C}-\boxed{OH+H}-NH(CH_2)_6NH_2 \xrightarrow[-H_2O]{} HO-\overset{\overset{\textstyle O}{\|}}{C}(CH_2)_8\overset{\overset{\textstyle O}{\|}}{C}-NH(CH_2)_6NH_2 \xrightarrow{\text{etc.}}$$

Sebacic acid   Hexamethylenadamine   (First step toward a Nylon 6-10 molecule)

Nylon 6-10 is only one member of the nylon family.[2] Another of these polyamides, Nylon 6-6, is prepared by condensing adipic acid and hexamethylenediamine. Segments of these polymers are

$$\left[-\overset{\overset{\textstyle O}{\|}}{C}(CH_2)_8\overset{\overset{\textstyle O}{\|}}{C}-NH(CH_2)_6NH-\right] \text{ and } \left[-\overset{\overset{\textstyle O}{\|}}{C}(CH_2)_4\overset{\overset{\textstyle O}{\|}}{C}-NH(CH_2)_6NH-\right]$$

Nylon 6–10   Nylon 6–6

The conditions for condensation of a dicarboxylic acid and a diamine are difficult to attain in the beginning chemistry laboratory, so you will synthesize Nylon 6-10 by a slight variation of the above procedure. Amides can be made easily, but more expensively, by heating an acid chloride,

with an amine. The first step in your synthesis of Nylon 6-10 from the acid chloride of sebacic acid (sebacoyl chloride) and hexamethylenediamine is

$$Cl-\overset{\overset{\textstyle O}{\|}}{C}(CH_2)_8\overset{\overset{\textstyle O}{\|}}{C}-Cl + NH_2(CH_2)_6NH_2 \xrightarrow[-HCl]{} Cl-\overset{\overset{\textstyle O}{\|}}{C}(CH_2)_8\overset{\overset{\textstyle O}{\|}}{C}NH(CH_2)_6NH_2 \xrightarrow{\text{etc.}}$$

Sebacoyl chloride

   The raw nylon that you will make is unsuitable for weaving into cloth without further refinement. Commercial nylon monofilament (thread) is made by forcing molten nylon through tiny orifices (openings). The thin streams of nylon harden into threads that are carefully stretched and twisted together to form a strong fiber.

_____

[1]An addition polymer results from covalent bond formation between monomers without the elimination of another molecule. The familiar polymer polyethylene is of the addition type.

$$nCH_2=CH_2 \longrightarrow [-CH_2-CH_2-]_n$$

Ethylene   A segment of a
polyethylene molecule

[2]The first number in the nylon name refers to the number of carbons in the diamine, the second to the number of carbons in the acid.

Hydrolysis of Cellulose

The glucose units (monomers) in a cellulose molecule are joined together by acetal bonds.

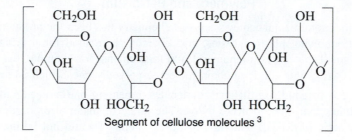

Segment of cellulose molecules [3]

Hydrolysis of these acetal bonds forms[3] β-glucose, which is rapidly converted to a mixture of all three forms of glucose— β-glucose, α-glucose, and the open-chain form.[4] In today's experiment you will attempt to detect these monomers.

---

[3]Carbons showing only three bonds also have a hydrogen.

[4]See experiment 26 for a discussion of the structures and reactions of glucose and starch. Also consult experiments 21 and 26 for a discussion of Fehling's test.

ENVIRONMENT, CULTURE, AND CHEMISTRY
Polymers and Petroleum

Certainly one of the monumental achievements of chemistry has been the production of polymers, found in applications using plastics, resins, and fibers. One method for making polymers is to condense (unite) small molecules (monomers) into long chains; for example, the condensation of simple ethylene, C_2H_4, into the long chain molecule polyethylene, which is a common plastic.

Another procedure is to condense different molecules, repeating one molecule after the other, as will be done in this experiment to make nylon. The range of material properties available from hundreds of polymeric (artificial) materials made by chemists could not be duplicated with the relatively few polymeric substances available in nature, such as starch, proteins, nucleic acids, and the building blocks of natural rubber. A majority of the technical achievement in medicine, communication, and manufacturing would simply vanish without the vast array of polymeric compounds. Imagine a world without Teflon, nylon, polypropylene, polystyrene, polyvinyl chloride (PVC), and many, many more. What is the raw material used in the production of the monomers to make these many different artificial polymers? The answer is the same as for so many of the products in our culture: petroleum and natural gas (methane). In specialized oil refineries, certain fractions of petroleum are converted by complex chemical reactions into numerous petrochemicals including monomers; probably the most important being ethylene, propylene, and benzene. Five percent of the crude oil used annually (200 million tons) is converted to petrochemicals. What will happen to the production of polymers when all of the petroleum has been consumed? We may be able to convert other fossil fuels, such as coal and oil shale, to fill this need, but with the end of the petroleum age, it is likely that plant materials (carbohydrates) may be used to make petrochemicals. For example, the stalks, cobs, and leaves that are waste products each year from corn production in the U.S. weigh about 100 million tons. At the present time there is much agricultural research with plants such as algae and with the natural prairie grass, Switch Grass, in which cellulose production is far less costly than corn. Who knows, perhaps Switch Grass, first encountered by the pioneers as they traveled west through the states of Illinois, Missouri, Kansas, and others, may become the polymer source for tomorrow.

Name _____

Date _____ Lab Section _____

Pre-Laboratory Questions | 27

1. Explain why the nylons are called polyamides.

2. Write equations for the reactions of benzoic acid and of benzoyl chloride,

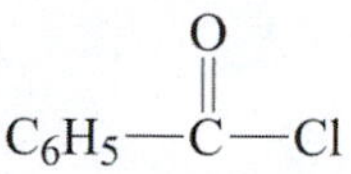

$$C_6H_5-\overset{\overset{\displaystyle O}{\|}}{C}-Cl$$

the acid chloride of benzoic acid, with *n*-propylamine to give an amide. Name the amide. In today's experiment, why will you synthesize nylon from an acid chloride rather than from a carboxylic acid?

3. Draw the structure of cellulose. Identify the acetal bonds, and circle monomer units.

4. A segment of a polymer molecule is shown below. Draw the structure of the monomer from which the polymer was synthesized. Is this an addition or a condensation polymer?

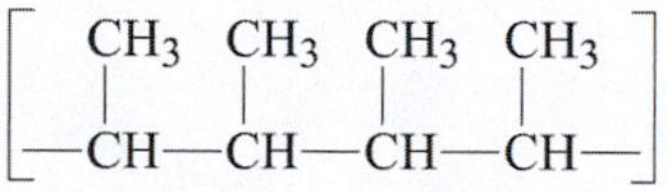

$$\begin{bmatrix} \overset{\displaystyle CH_3}{|} & \overset{\displaystyle CH_3}{|} & \overset{\displaystyle CH_3}{|} & \overset{\displaystyle CH_3}{|} \\ -CH- & CH- & CH- & CH- \end{bmatrix}$$

EXPERIMENT

1. Hydrolysis of a Polymer, Cellulose: Formation of a Monomer

Suggestion: Because of the danger of sulfuric acid, your instructor may choose to demonstrate this part of the experiment. This part of the experiment should be started at the beginning of the laboratory session. Make a compact wad of cotton (must be natural cotton) by compressing a cotton ball (about 4 cm in diameter), and place it in a large mortar. Pour 3 mL of concentrated sulfuric acid, H_2SO_4, over the cotton and grind it to a fine pulp with the pestle. Cautiously add 50 mL of water to the pulp, stir, and transfer all of the material to a 250-mL beaker. Boil the cellulose solution *very gently* for 30 minutes. Replace any water that evaporates during the boiling process.

When the hydrolysis solution has cooled, neutralize it by stirring in 20% sodium hydroxide, NaOH, solution until the solution is just basic to litmus. Confirm the presence of the monomer in the aqueous solution by means of Fehling's test as follows: Mix 1 mL of the hydrolysis solution with 2 mL of Fehling's reagent (equal amounts of solutions A and B) and heat in a water bath for 10 to 15 minutes. Formation of a reddish brown precipitate of cuprous oxide, Cu_2O, constitutes a positive test for the monomer.[5] (Answer Question 1, a–d on the Observations and Results sheet.)

> **Caution:** Concentrated H_2SO_4 is dangerous! If you come in contact with acid, wash it off immediately.

2. Synthesis of a Polymer, Nylon

Do this part of the experiment in teams of two. Do not get any of the reagents on your skin, since they are highly irritating. Clean and dry a 50-mL beaker. To the beaker add 10 mL of a 1.5% solution of sebacoyl chloride in 50:50 hexane/methylene chloride. Obtain 5 mL of a solution that contains 4% hexamethylenediamine and 3% sodium hydroxide.

Tilt the beaker containing the sebacoyl chloride-hexane/methylene chloride solution, and carefully pour the hexamethylenediamine solution down the side of the tilted beaker in such a manner that the aqueous solution forms a layer on top of the denser hexane/methylene chloride solution; *the layers must not mix together!* Put a piece of paper towel on the floor, and place the beaker on the towel.

Notice that the monomers have reacted at the interface between the layers to form a film of nylon. Use your spatula to loosen the film from the sides of the beaker. Now grasp the nylon film in the center with a wire hook (figure 27.1) and slowly lift it, twisting as you lift. As you continue to lift and twist, a rope of nylon will form. Pull out about 1 m of rope and cut it with a pair of scissors. Put a piece of paper towel under the rope so that the chemicals do not drop on the desk or floor and carry it to the hood. Dip the rope into a 50% aqueous alcohol solution, and thoroughly rinse it to remove all of the chemical reagents. Stretch out the nylon rope on a paper towel under the hood and press it with paper towels. When the rope is dry, cut off 30 cm of it, label it with your names (use masking tape for a label), and submit it to your instructor. Cut the remaining rope into short lengths (5 to 10 cm) for tests described in the next paragraph.

---

[5]See experiment 26 for a discussion of the structures and reactions of glucose and starch. Also consult experiments 21 and 26 for a discussion of Fehling's test.

Caution: Do this experiment with adequate ventilation or in a fume hood. Do not touch the polymer with your hands until it has been washed.

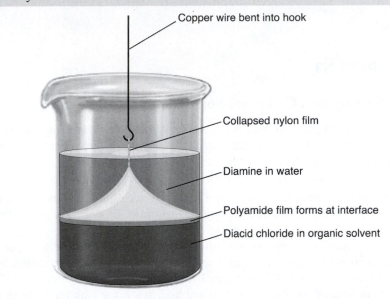

▲ **FIGURE 27.1** Synthesis of nylon at the interface between a diacid chloride and a diamine.

Test the strength of your nylon by pulling on both ends of a piece. Using forceps, hold a piece of nylon in a flame to see if it will burn. Determine whether your nylon pieces are soluble in, or are affected by, the following solvents: acetone (under the hood), commercial liquid bleach, 25% sulfuric acid, 25% sodium hydroxide, and concentrated sulfuric acid (a small beaker of sulfuric acid will be provided under the hood—dip a piece of fiber into the solution). Now repeat all of these tests on cotton and compare the results with those for nylon. You may wish to pull out some more nylon rope to take with you to show your friends. Be sure to wash it carefully as you did before. (Answer Question 2, A–D.)

Destroy the remaining monomers by stirring the two layers together with your spatula or stirring rod. Put the ball of nylon as directed by your instructor (*never* down the drain), and dispose of the remaining solvents

QUESTIONS AND PROBLEMS

1. Is amylose an example of an addition or a condensation polymer? Explain.

2. Draw the structures of cellulose and amylose, side by side, and discuss the difference in their structures. Do both of these polysaccharides give the same product upon hydrolysis? Explain.

3. Write an equation for the synthesis of Nylon 6-6 from adipoyl chloride and hexamethylene-diamine.

4. Draw the structure of a segment of the nylon that would be obtained by heating oxalic acid with ethylenediamine, $NH_2CH_2CH_2NH_2$. Name this nylon, using the numbering system discussed in this experiment.

5. What raw material is used to make synthetic plastic and polymers? What raw material might be used in the future to make polymers in an economy operated on a sustainable basis?

6. List at least six commercial uses of nylon. For which of these uses would cotton also serve?

Name _____

Date _____ Lab Section _____

EXPERIMENT

Natural and Synthetic Polymers: The Preparation of Nylon

27

OBSERVATIONS AND RESULTS

1. Hydrolysis of a Polymer, Cellulose: Formation of a Monomer

a. Describe the results of Fehling's test.

b. What is the name of the monomer that is identified by this test? Draw the structure of the monomer that is present in the hydrolysis solution.

c. Write an equation for the reaction that occurs between the monomer and Fehling's reagent.

d. Draw a segment of a polymer (cellulose) molecule, and show how water molecules break the acetal bonds to form a monomer molecule.

2. Synthesis of a Polymer, Nylon

A. Draw the structure of each monomer, and circle the functional groups that are involved in the polymerization (condensation) reaction.

B. Using an appropriate functional group from each monomer, show the condensation step.

C. Describe the results of the various tests on your nylon.

1. Strength:

2. Flammability:

3. Acetone:

4. Sulfuric acid (25%):

5. Sulfuric acid (concentrated):

6. Sodium hydroxide (25%):

7. Commercial liquid bleach:

D. If you had heated your nylon in aqueous sulfuric acid (25%) for an hour or so, it would have reacted and dissolved. Write an equation for the reaction that takes place under those conditions.

Triglycerides and
Other Lipids | 28

OBJECTIVES

1. To observe the physical properties of some common lipids.

2. To make quantitative comparisons of the extent of unsaturation of some fats.

3. To establish the relationship between the structures of the fatty acid residue in the lipid and the chemical and physical properties of the lipid.

DISCUSSION

One topic of conversation that invariably comes up in the chatter at social events is that of diet. There are enough diets and approaches to dieting to confuse the mind of a Solomon. Most diets have an immediate goal—weight loss. So-called excess weight is primarily fat or, in chemical terminology, triglyceride. Triglycerides are members of the lipid family.

Not only are triglycerides central to the problem of obesity, but certain of them, along with another lipid, cholesterol, are suspect in arteriosclerosis, "hardening of the arteries." However, triglycerides are not, by any means, all bad. They serve as important sources of energy in foods such as butter, margarine, vegetable oils, shortening, meat products, and peanut butter.

Triglycerides also fulfill other essential dietary functions. They serve as carriers for the fat-soluble vitamins—vitamins A, D, E, and K. Elimination of fat from the diet causes reduction in intake of these nutrients. Certain triglycerides serve as the sole source of the unsaturated fatty acid linoleic acid. Since this compound cannot be synthesized by the body, linoleic acid is considered an essential fatty acid. Finally, triglycerides delay the feeling of hunger following a meal because they leave the stomach slowly.

In today's experiment you will become acquainted with several triglycerides, as well as other members of the lipid family. Included in this family are a wide assortment of biological substances. Lipids are distinguished from carbohydrates, proteins, and nucleic acids by their solubility in nonpolar organic solvents.

Structure of the Triglycerides and Fatty Acids

Triglycerides, called fats and oils, are esters in which the "alcohol portion" of the molecule is always glycerol and the "acid portion" generally is derived from long-chain acids, called fatty acids. The expression *fatty acid residue* will be used to refer to the fatty acids in the triglyceride molecule.

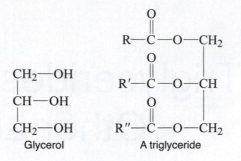

Glycerol   A triglyceride

As the R–groups imply, there may be as many as three different fatty acid residues within a triglyceride molecule; hence the name *mixed glyceride*. Since the fatty acid residues vary somewhat from one molecule to another, each particular fat or oil (cottonseed oil, butter, etc.) is a mixture. Some common fatty acids incorporated in triglycerides are shown below.

$$CH_3(CH_2)_{14}-\overset{\overset{\displaystyle O}{\|}}{C}-OH \qquad CH_3(CH_2)_7-CH=CH-(CH_2)_7-\overset{\overset{\displaystyle O}{\|}}{C}-OH$$

Palmitic acid   Oleic acid

$$CH_3(CH_2)_4-CH=CH-CH_2-CH=CH-(CH_2)_7-\overset{\overset{\displaystyle O}{\|}}{C}-OH$$

Linoleic acid

You will note that palmitic acid is saturated, whereas the other two acids are unsaturated. A triglyceride derived from glycerol and these three fatty acids is typical of a molecule in cottonseed oil.

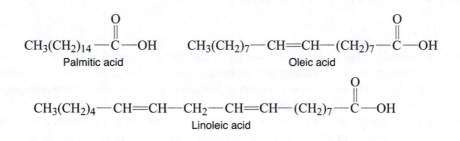

A typical molecule in the cottonseed oil

Other important saturated acids are lauric, myristic, and stearic.

$$C_{11}H_{23}-\overset{\overset{\displaystyle O}{\|}}{C}-OH \qquad C_{13}H_{27}-\overset{\overset{\displaystyle O}{\|}}{C}-OH \qquad C_{17}H_{35}-\overset{\overset{\displaystyle O}{\|}}{C}-OH$$

Lauric acid   Myristic acid   Stearic acid

Unsaturated acids are linolenic (with three double bonds) and arachidonic (with four double bonds).

$$C_{17}H_{29}-\overset{\overset{\displaystyle O}{\|}}{C}-OH \qquad C_{19}H_{31}-\overset{\overset{\displaystyle O}{\|}}{C}-OH$$
Linolenic acid   Arachidonic acid

Fats and Oils: Solids and Liquids

The saturated fatty acids are solids at room temperature, whereas the unsaturated acids are liquids. The same general relationship between unsaturation and melting points holds for the triglycerides: the melting points decrease as the proportion of unsaturated fatty acid residues in the triglyceride increases. We find, therefore, that a triglyceride such as tallow, which is rich in saturated fatty acid residues (approximately 90%), is a solid. On the other hand, olive oil, containing approximately 86% oleic and linoleic acid residues, is a liquid. We can conclude that the physical state of the fat, whether liquid or solid, provides a rough idea of the type of fatty acid residues that are present. It has become customary to call solid triglycerides fats and liquid ones oils.

Other Lipids

Triglycerides are the most common members of the lipid family, but they are not the only important ones. Others are the steroids and the lecithins. Still another, lanolin (used as a base for salves and ointments), is a complex mixture of esters, steroids, hydrocarbons, and other compounds.

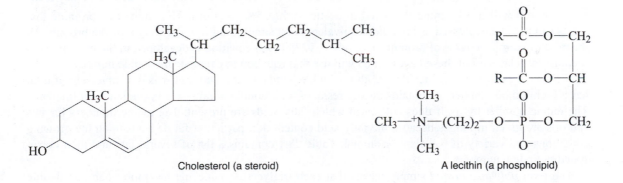

Cholesterol (a steroid)   A lecithin (a phospholipid)

Reactions of the Triglycerides

The two principal points of reaction in a triglyceride molecule are at the ester and alkene (if one exists) functional groups. The important reactions of the ester functional group are hydrolysis and saponification. You will study saponification in the next experiment. The reactions that occur at the alkene functional group are hydrogenation (addition of hydrogen, H_2), development of rancidity, "drying oil" reactions, hydration (addition of water), and halogenation (addition of halogen); it is the last reaction (addition of halogen) that you will use in today's experiment.

The halogenation reaction involves the addition of a halogen such as chlorine, Cl_2, bromine, Br_2, or iodine, I_2, to an alkene. (See Experiment 19 for a review of the addition of bromine to alkenes.) Reaction of bromine with a triglyceride typical of lard is shown below.

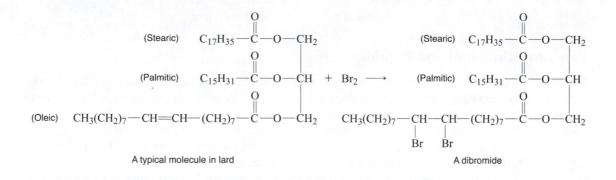

A typical molecule in lard   A dibromide

Bromination reactions will be used to establish the extent of unsaturation of fats.

The Iodine Number

A convenient measure of the degree of unsaturation is given by the iodine number. We define the iodine number of a fat as the weight of iodine that reacts with 100 g of the fat. Because the reaction with iodine is very slow and often incomplete, actual measurement is made with a more reactive reagent such as BrI, BrCl, or Br_2, and the weight of reagent absorbed is converted to the iodine equivalent. In today's experiment you will use bromine dissolved in acetic acid (a 5% solution). The amount of bromine that reacts can then be expressed as an iodine equivalent. You simply multiply the weight of the bromine by the ratio of atomic weights of bromine and iodine, 127/80. An equation is developed in the experimental section that relates all of these factors. You will use that equation to calculate the iodine number.

If a fat contains fatty acid residues with no double bonds, its iodine number is 0. Conversely, if a fat has a high iodine number, it contains mostly residues with double bonds—oleic, linoleic, and linolenic. The iodine number by itself does not reveal which fatty acids are present. The iodine numbers for fats and oils are given in ranges because the fatty acid content of a particular fat depends upon the growing conditions and variety of the plant or animal. Table 28.1 contains a list of several fats and their corresponding iodine numbers.

The term *polyunsaturated* simply means that each triglyceride molecule has more than one double bond. A correlation has been observed between diets rich in saturated fats and the disease arteriosclerosis. Hence, some in the medical profession recommend the use of plant oils (which are relatively unsaturated) rather than animal fats (which are relatively saturated).

TABLE 28.1 Reported Iodine Numbers of Some Common Fats and Oils

Fat or Oil	Iodine Number	Fat or Oil	Iodine Number
Coconut	8–10	Olive	75–95
Butter	50–60	Peanut	84–100
Tallow	25–45	Corn	115–130
Lard	30–48	Cottonseed	105–115
Palm	45–65	Soybean	125–135
Sardine	120–190	Safflower	130–150

ENVIRONMENT, CULTURE, AND CHEMISTRY
Saturated, Unsaturated, and *Trans* Lipids (Fats): Heart Disease and Orangutans

Lipids (fats) are part of our food supply, both in plants and animals, as esters with glycerol as the alcohol portion of the lipid molecule. The acid part (called a fatty acid) consists of long carbon chains, both saturated and unsaturated. The most common saturated acids (alkane-like) are stearic, palmitic, and myristic; the common unsaturated ones (alkene-like) are oleic, linoleic, linolenic, and arachidonic. The saturation/unsaturation ratio of a lipid primarily determines whether it will be a solid or a liquid (oil); common solid (saturated) fats are butter, lard, and beef tallow; peanut, olive, and palm (unsaturated) are oils. All naturally occurring unsaturated fats have one or more carbon-carbon double bonds (alkenes) in a *cis* configuration, and are healthier in the diet since they do not lead to the buildup of cholesterol and contribute to coronary heart disease (CHD). About one hundred years ago, it was discovered that partial hydrogenation of the lipid oils (conversion of alkenes to alkanes with H_2 and a catalyst) led to a butter-like solid that was easier to spread on bread than butter, slower to decompose at room temperature (less necessity for refrigeration), and easier to use in baking. Margarine was the name given to this solid. Margarine and baking shortening, also made by partial hydrogenation, became extremely popular; some thought that margarine might displace butter as a table spread. Then, as is so common with the inventions of science, in the 1990s a cloud appeared on that horizon; *trans* fats, with a direct, serious link to increased CHD, were detected at a significant level in the partially hydrogenated oils; apparently the hydrogenation process inverted the double bond from the *cis* to *trans* configuration. A definitive study, completed in 2006 confirmed the connection between *trans* lipids (fats) and CHD. Now, because of changes in the hydrogenation process, it is possible to buy margarine without *trans* fat. Recently palm oil, obtained by crushing the fruit of a palm native to Malaysia and Indonesia, has become extremely popular for cooking, baking, biofuel, and many other applications; palm oil has approximately a 50:50 ratio of saturated to unsaturated fat without any *trans* fat. Even the saturated fatty acids in palm oil are safer to eat because the carbon chains are shorter. However, there is a serious downside to this story too. Approximately 35 million acres of rain forest in Malaysia and Indonesia have been cleared to raise palms for their oil; currently this amount is growing by about 4.6 million acres a year. The rapid and increasing growth in palm oil plantations has severely reduced the habitat of the orangutan, rapidly driving this curious and elusive higher primate toward extinction. Over the past thirty years, some 3,000 orangutans have died per year. The question is: Can orangutans and palm oil coexist?

Pre-Laboratory Questions | 28

1. Is the "alcohol portion" of all triglycerides the same? Explain.

2. What is a fatty acid residue?

3. Write complete structures for the following fatty acids: myristic, linolenic, and lauric. Which acid is unsaturated?

4. What is the difference, if any, between a fat and an oil?

5. Which fat or oil in table 28.1 is most unsaturated?

EXPERIMENT

1. Observation of Some Lipids and Fatty Acids

Observe the appearance and consistency, and waft the following lipids and fatty acids: cholesterol, lecithin, palm oil or coconut oil, beef fat (tallow), corn oil, cottonseed oil, stearic acid, oleic acid, and lanolin. (Answer Question 1, a–g on the Observation and Results sheet.)

2. Determination of Iodine Numbers of Fats and Oils

The purpose of this part of the experiment is to compare the degree of unsaturation of some common household fats by determining their iodine numbers. You will compare a commercial cooking oil, a margarine, and an animal fat or oil, such as butter. Work in groups of two. One student should use an oil and a margarine that are advertised as being high in polyunsaturates, and the partner should use an oil and a margarine that are not so advertised. To determine the iodine number you will add bromine, dissolved in acetic acid, from a weighed reagent flask to a fat or an oil, dissolved in methylene chloride, CH_2Cl_2, in a reaction flask, until the bromine color in the reaction flask matches the color of the bromine solution in the blank flask.[1]

A. Preparation of the Reagent Flask

Prepare a bromine reagent flask from a 25-mL Erlenmeyer flask. Fit the flask with a dropping pipette by means of a one-hole cork or stopper (figure 28.1). Fill the flask three-quarters full with the 5% (by weight) bromine–acetic acid solution.

> **Caution:** Do not get the bromine–acetic acid solution on your skin. If this happens, wash it off immediately, since it will cause severe burns.

B. Preparation of the Reaction Flask

Now add 0.2 to 0.3 g (precisely weighed) of your fat or oil to the tared (zeroed) reaction flask (a second 25-mL Erlenmeyer flask). Liquids may be added with a dropper and solids with a wooden splint or spatula. Attempt to deliver the solids to the bottom of the reaction flask so that they will dissolve readily. Dissolve all of the fat or oil in 5 mL of methylene chloride, CH_2Cl_2, by stirring it in with a glass stirring rod.

C. Preparation of the Blank Flask

A third 25-mL Erlenmeyer flask will serve as an end-point blank. Put into this flask 10 mL of methylene chloride, CH_2Cl_2, and 10 drops of the bromine–acetic acid solution. Stopper the flask. Place the blank flask and the reaction flask side by side on white paper.

---

[1]Since the endpoint (the appearance of brown color) is not sharp, it is more accurate to add bromine slightly past the endpoint until the color exactly matches the color of a blank solution whose concentration is known. The weight of bromine in the blank must always be subtracted from the total weight of bromine to obtain the weight of bromine that reacted with the fat.

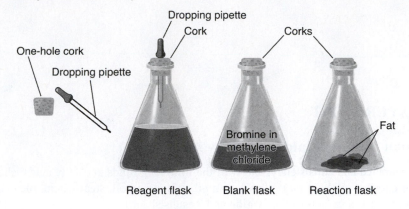

▲ **FIGURE 28.1** Equipment for determining the iodine number.

D. Determination of the Iodine Numbers

Now you are ready to determine the iodine numbers. Weigh the *reagent flask* to two decimal places and record the weight. At first, add bromine solution rapidly to the *reaction flask* (squirt in about 10 drops at a time *and swirl the flask after each addition*). Add the bromine drop by drop as you approach the end point. Continue until the color in the *reaction flask* matches the color of the *blank*. The color should persist for about half a minute. Weigh the *reagent flask* again and record the weight. The difference between the initial and final weights of the *reagent flask* equals the weight of bromine solution used in the reaction.

Dispose of waste from the reaction flask for each determination and wash the flask with water.

Shake as much water from the reaction flask as possible and dry the flask with a piece of paper towel before weighing another sample of fat or oil into it. Repeat the determination. Do the procedure a third time if the results from the first two trials are not close.

An equation[2] for determination of the iodine number is

$$\text{Iodine number} = \frac{(\text{grams of bromine solution} - 0.2 \text{ g}) \times 0.05 \times \dfrac{127}{80} \times 100}{\text{grams of fat}}$$

Simplification gives

$$\text{Iodine number} = \frac{(\text{grams of bromine solution} - 0.2 \text{ g}) \times 7.94}{\text{grams of fat}}$$

(Calculate the Iodine numbers in 2D, and answer Question 2D, 1–6.)

---

[2]To explain the terms of the equation: 0.2 g (10 drops) of excess bromine solution was added in each determination to match the intensity of the blank; the factor 127/80 converts "bromine number" to iodine number; a 5% (0.05) solution of bromine in acetic acid was used; and it is necessary to multiply by 100 to convert the result to 100 g of fat.

QUESTIONS AND PROBLEMS

1. As you noted in your experiment, several manufacturers claim that their fat or oil is more unsaturated than a competitor's product. Why is there concern about the amount of saturated fat in our diet?

2. Draw the structure of a triglyceride molecule that contains the following fatty acid residues: myristic acid, oleic acid, and linolenic acid. What is the iodine number of this triglyceride?

3. Using the triglyceride of question 3, which is an oil, write a reaction showing the conversion of this oil to a solid shortening. What is the name of this type of reaction?

4. One triglyceride contains two stearic acid residues and one linoleic acid residue; another has two oleic acid residues and one stearic acid residue. Would the triglycerides have identical iodine numbers?

5. What caused the appearance of trans fats in some fat products? What is the difference between cis and *trans* fatty acids?

EXPERIMENT

Triglycerides and Other Lipids | 28

OBSERVATIONS AND RESULTS

1. Observation of Some Lipids and Fatty Acids

   a. Which of the compounds examined is a pure steroid?

   b. Which has the most pleasant odor? The least odor?

   c. Which of the compounds is a phospholipid?

   d. Give the names of the lipids that are mixtures.

   e. Are any of the *lipids* pure compounds? If so, give their names.

   f. Based on *your observations of the compounds*, which has the higher melting point, stearic acid or oleic acid? What is your evidence?

   g. Give the names of the lipids that are primarily triglycerides, and indicate which of these triglycerides contains the most saturated fatty acid residues.

2. Determination of the Iodine Numbers of Fats and Oils

Name of partner _____

D. Name of Fat or Oil	Weight (g) of Fat or Oil		Weight (g) of Reagent Flask and Bromine Solution			Iodine Number	
	Run 1	Run 2		Run 1	Run 2	Run 1	Run 2
_____	____ ____		Initial	____	____		
			Final	____	____		
			Solution used (grams)	____	____	____ ____	
						Average ____	

Name of Fat or Oil	Weight (g) of Fat or Oil		Weight (g) of Reagent Flask and Bromine Solution			Iodine Number	
	Run 1	Run 2		Run 1	Run 2	Run 1	Run 2
_____	____ ____		Initial	____	____		
			Final	____	____		
			Solution used (grams)	____	____	____ ____	
						Average ____	
Butter	____ ____		Initial	____	____		
			Final	____	____		
			Solution used (grams)	____	____	____ ____	
						Average ____	

1. Which of the cooking oils (or margarines) did you find to be most unsaturated? Do your data agree with the manufacturer's claim? Explain.

2. Examine its container and name the vegetable or animal source from which the triglycerides in your cooking oil (or margarine) were obtained. If the iodine numbers for either of your cooking oils are listed in table 28.1, compare these values to your experimentally determined values.

3. Based on the data that you obtained in this experiment and on information in the discussion section, can you make the statement that vegetable fats are more unsaturated than animal fats? Explain.

4. Compare your iodine number for butter with the reported value in table 28.1. Are the values quite similar?

5. According to your experimental data, which contains the most unsaturated fatty acid residues, a molecule of butter, a molecule of cooking oil, or a molecule of margarine? Explain.

6. Fatty acid residues that are typical of a butter molecule are myristic, palmitic, and oleic. Draw the structure of a molecule of this triglyceride, and show its reaction with bromine.

Saponification: Soaps and Detergents | 29

OBJECTIVES

1. To carry out and make observations on the saponification of a triglyceride (a fat).
2. To prepare a soap and examine its properties.
3. To isolate a mixture of fatty acids obtained by acidification of a soap solution.
4. To gain an understanding of the cleansing action of soaps and detergents in hard and soft water.

DISCUSSION

Few college students today can recall watching soap being made. Only a few generations ago people routinely made soap by boiling beef tallow (a triglyceride) with lye (impure sodium hydroxide). When the top layer cooled and solidified, it was cut into cubes of yellow soap (sodium stearate). The soap may have been hard on hands and clothes, but it had good cleaning action, particularly in soft water. In this experiment you will make soap much as your grandmother may have and examine some of its properties.

Saponification[1] of Triglycerides

Triglycerides are high-molecular-weight fats and oils (esters) that can be saponified (hydrolyzed) in basic solution to give soap and glycerol.

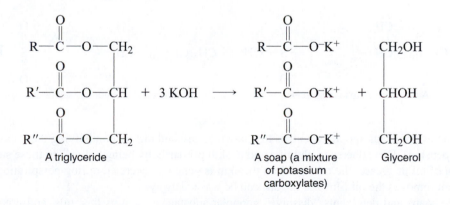

[1]Saponification comes from the Latin word *sapon*, meaning soap. For a discussion of the saponification of esters, see Experiment 22.

As suggested in the equation, soap is a salt composed of a mixture of carboxylate anions and a univalent cation. A mixture of anions is formed because each triglyceride molecule contains a variety of fatty acid residues and because a particular fat or oil is itself a mixture of molecules.[2]

Potassium soaps are more soluble than sodium soaps and readily produce a lather. Therefore, potassium soaps are used to make liquid soap and shaving cream. Soaps from highly saturated, solid fats, such as tallow, lard, or shortening, are hard. Saponification of an unsaturated oil, such as olive oil, gives a soft soap.

Treatment of a soap solution with dilute hydrochloric acid produces a mixture of fatty acids.

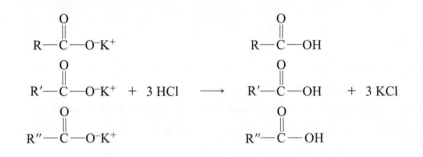

Fatty acids are long-chain carboxylic acids (C_{10} to C_{18}) that may be saturated or unsaturated (see Experiment 28).

Synthetic detergents differ from soaps in that they are salts of long-chain alkyl sulfuric acids or alkylbenzenesulfonic acids, rather than carboxylic acids.

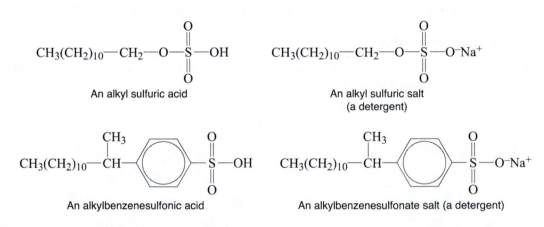

An alkyl sulfuric acid

An alkyl sulfuric salt
(a detergent)

An alkylbenzenesulfonic acid

An alkylbenzenesulfonate salt (a detergent)

The function of soaps and detergents is to remove grease and dirt by emulsifying the grease (bringing it into suspension). Dirt adheres to clothing and to skin primarily by being "glued" to these surfaces with a thin film of oil or grease; the oil (lipid) on the skin is generally secreted during perspiration. The soap or detergent removes the oil film and the dirt can be washed away.

How do soaps and detergents "dissolve" nonpolar substances such as fats, oils, and greases? Molecules of soaps and detergents contain a nonpolar hydrocarbon end and a polar end, which is usually ionic. The nonpolar ends of the molecules surround the tiny oil droplets and are partially dissolved in them (like dissolves like). The polar ends of the molecules, which are extremely soluble in water, solubilize or emulsify the entire droplet (figure 29.1).

---

[2]For a more complete discussion on the structures of triglycerides and fatty acids, see Experiment 28.

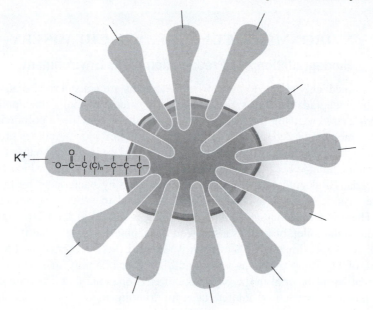

▲ **FIGURE 29.1** A soap micelle. Soap molecules surround the oil droplet and stick their hydrocarbon ends into it, thus making the entire micelle water soluble.

Soaps do not work well in hard water, because the divalent cations of dissolved minerals (Ca^{2+}, Mg^{2+}, and Fe^{2+}) form precipitates with the carboxylate (fatty acid) anions of soaps. Consequently, a scum of calcium stearate (and other salts) is typically found as a ring in the bathtub or as a dingy film on clothes, hair, or skin.

$$CH_3(CH_2)_{16} - \overset{\overset{\displaystyle O}{\|}}{C} - O^-Ca^{2+-}O - \overset{\overset{\displaystyle O}{\|}}{C} - (CH_2)_{16}CH_3$$

Calcium stearate (a scum in hard water)

On the other hand, the alkyl sulfate and alkyl sulfonate anions of detergents do not form precipitates with these cations, so they are quite effective in hard water.

In today's experiment you will prepare a potassium soap by saponification of a fat with potassium hydroxide. You will use the potassium soap to make a sodium soap and to obtain a mixture of fatty acids. You will also compare some of the properties of soaps and detergents.

ENVIRONMENT, CULTURE, AND CHEMISTRY
Biodegradation and Preservation of the Environment

Biodegradation immediately comes to mind when detergents are considered because the early detergents were nonbiodegradable, that is, they persisted in nature for a long time, perhaps for years. A memory of a televised segment from the 1960s brings this back to life: a polluted river in northern Ohio showed mounds of detergent foam, like great drifts of snow. Another problem with detergents is the high percentage of phosphates, which lead to rapid growth of algae. Today, detergents are biodegradable because of modifications in the carbon chain, which make them susceptible to bacteria. Biodegradation is usually accomplished by bacteria or by hydrolysis (reaction) with water. Biodegradation and recycling are essential to controlling society's waste and contaminants. There is a tendency to think if something is placed in the trash, washed down the drain, flushed down the toilet, or used as a spray on trees, plants, or grass that it is gone. The facts are not that simple. Approximately 32% of trash is recycled, 12% is burned, and 56% is buried in landfills. A few of the many problems with landfills are: availability of suitable land, non- (or slow) biodegradation of material in the oxygen-poor environment, slow leaching of toxic chemicals, including medical wastes, into aquifers used for drinking water, and formation of flammable methane. Most urban centers in advanced countries have centers for purifying wastewater. Treatment consists of collecting the wastewater in a central location followed by filtration, time for biodegradation to occur, and other procedures. This process results in water that is often nearly pure; in some cases it is drinkable. Wastewater purification is often inadequate in the cities of many of the less advanced nations, and frequently the wastewater is poured directly into streams, rivers, or the ocean, leading to extensive pollution and health hazards. Use of insecticides and herbicides (weed killers) also requires biodegradation, or toxicity may become a problem. Consider an insecticide sprayed on fruit: the chemical must be degraded before the fruit can be eaten. One of the first insecticides, DDT, illustrates the problem of biodegradation. Developed shortly after World War II, DDT was quickly recognized for its potency against a wide range of damaging insects. Soon, however, it was determined that DDT had spread widely from its original point of application, even appearing in the fat tissue of the penguins in the Antarctic. Then the brown pelicans in California began to disappear. Scientists established that DDT was the culprit. Present in the fish eaten by the pelicans, it caused their eggshells to be soft so they were crushed by the mother bird before they could hatch. The problem with DDT (banned by most countries) is its long lifetime of 2–15 years (essentially nonbiodegradable), giving time for its toxicity to become a serious problem. DDT has been banned by most countries, but this has led to an increase in mosquitoes and the accompanying problem of West Nile disease, at least in the U.S. The goal of the chemical companies with the more recent insecticides and herbicides is to increase their rate of biodegradation. Two common insecticides, Malathion and Carbaryl (Sevin), biodegrade in about two weeks or less; the chemical in the very popular herbicide, Roundup, lasts longer, even up to ninety days on the leaves. Roundup has toxicity problems, and some states have required that "biodegradable" be removed from its label. It is clear that in order for the environment to be sustained, biodegradation steps must be built into most new products and procedures.

Pre-Laboratory Questions | 29

1. Draw a molecule of a typical solid fat and write an equation for its saponification to a sodium soap.

2. Draw the complete structures, showing all bonds, for stearic acid and sodium stearate.

3. In what ways does a sodium soap differ from a potassium soap?

4. Draw structures for carboxylate ion, alkyl sulfate ion, and alkylbenzenesulfonate ion.

5. Explain why oils and fats (lipids) and greases (hydrocarbons) are insoluble in water.

6. Write the structure for calcium stearate. Is this salt soluble in water?

 **EXPERIMENT**

1. Saponification of a Fat: Preparation of a Potassium Soap

Weigh about 1.5 g of solid fat (tallow, lard, or shortening) in a *large* test tube. (It is not necessary to force the fat to the bottom of the test tube, since it will melt and run down when the test tube is heated.) Add 10 mL of a 10% solution of KOH in 95% ethyl alcohol.[3] Place the test tube in a 250-mL beaker half filled with hot water and continue boiling with a hot plate.[4] Replace any alcohol that evaporates with ethyl alcohol. **Caution:** *Alcohol is flammable.*

After heating the tube for 15 minutes, test for completeness of saponification by adding a few drops of the reaction mixture to water. Do you see droplets of fat? If not, saponification is complete. If droplets of fat are visible, continue to heat the tube for another 15 minutes and test again. When saponification of the fat is complete, pour the contents of the test tube into a 100-mL beaker. Heat the beaker on a hot plate until the alcohol is gone and the residue becomes viscous and tacky; do not overheat it or the soap will darken. Now add 30 mL of distilled water, and heat the mixture briefly, while stirring it, until a solution is obtained. The product is a solution of a potassium soap. Divide the solution into two parts. Prepare a sodium soap from one half and retain the other half for tests. (Answer Question 1, a–d on the Observations and Results sheet.)

2. Formation of a Sodium Soap

To half (save the other half) of the potassium soap solution add 15 mL of a saturated sodium chloride solution (made with distilled water). Stir vigorously until large curds appear. Isolate the solid by pouring the mixture through filter paper in a funnel. Press the soap between several pieces of paper towel to remove the water. You will use some of your sodium soap in another part of this experiment. Turn in a small piece to your instructor at the end of the experiment. You may keep the rest, unless instructed otherwise. (Answer Question 2, a–c.)

3. Preparation of Fatty Acids from a Soap

Dilute your remaining potassium soap solution with an additional 30 mL of distilled water. Put 5 mL of this diluted soap solution in a test tube and add dilute hydrochloric acid, drop by drop, until the solution is acidic to litmus. Shake the mixture to test for sudsing action. Remove the solid with a stirring rod and determine whether it is soluble in acetone by adding a small piece to 1 mL of acetone under the hood. Now determine whether a small piece of the original fat and the sodium soap are soluble in 1 mL of acetone. (Answer Question 3, a–e.)

> **Caution:** Acetone is extremely flammable and must be kept away from burners!

---

[3] Alcohol, rather than water, is used because it dissolves both the base and the fat.

[4] If hot plates are not available, you may heat the beaker (supported on a wire gauze) with a burner. If the alcohol ignites, extinguish the flame with a towel or with the bottom of another beaker.

4. Properties of Soaps and Detergents

Test a 1-mL portion of your potassium soap solution for a sudsing action. Smear a tiny piece of the starting fat on the bottom of a watch glass, and determine whether your potassium soap solution removes the grease. Perform the same tests with a little of your sodium soap added to water.

To 1-mL portions of your potassium soap solution in three separate test tubes, add 1 mL of each of the following solutions: calcium chloride (0.1%), magnesium chloride (0.1%), and ferrous chloride (0.1%). Stir the solutions. Do precipitates form? Repeat this test with a synthetic detergent such as Tide. Prepare a solution of the detergent by dissolving 0.5 g of detergent in 50 mL of distilled water.

Now mix equal portions (1 mL each) of your potassium soap solution and tap water. Does a precipitate form? Repeat with the detergent solution. (Answer Question 4, a–f.)

> **Dispose of all waste in the appropriate container.**

QUESTIONS AND PROBLEMS

1. The structure of a sodium sulfonate is shown in the discussion section of this experiment. Draw the structure of the calcium salt of this detergent. Is this salt soluble in water?

2. A water softener is an ion-exchange unit that replaces calcium, magnesium, and ferrous ions with sodium ions. How is a water softener important in increasing the effectiveness of soap? Does a water softener convert "hard water" to "soft water"? Explain.

3. Does dry cleaning involve detergents and soaps? What kinds of chemicals are involved in dry cleaning, and how do they function?

4. Describe how you could make a hard soap, starting with olive oil.

5. Suppose that the lipid that you used in today's experiment was a mixture of a triglyceride and a cholesterol. How could you use the saponification process to separate the cholesterol from the triglyceride?

EXPERIMENT

Saponification: Soaps and Detergents

29

OBSERVATIONS AND RESULTS

1. Saponification of a Fat: Preparation of a Potassium Salt

a. Draw the structure of a triglyceride that is representative of the fat used in today's experiment and write an equation for the saponification.

b. Explain how addition of the reaction mixture to water can serve as a test for completeness of saponification.

c. Explain on the basis of structure why the products of your saponification reaction are soluble in water whereas the starting fat is insoluble.

d. What happened to the glycerol that was formed during your saponification reaction?

2. Formation of a Sodium Soap

   a. Write an equation for the conversion of your potassium soap (use a typical fatty acid residue) to a sodium soap, showing the structures of reactants and products.

   b. Describe the texture of your sodium soap.

   c. Is the sodium soap a salt? Explain.

3. Preparation of Fatty Acids from a Soap

   a. Write an equation for the reaction of the potassium soap (a typical molecule) with hydrochloric acid, showing the structures of reactants and products.

   b. Did the solution show sudsing action after addition of acid? Explain the behavior.

   c. Describe the texture of your fatty acid preparation.

d. Did you isolate a single type of fatty acid, or a mixture of fatty acids? Explain.

e. Were the fatty acids soluble in acetone? Was the fat soluble? Was the soap soluble?

4. Properties of Soaps and Detergents

a. Did your sodium and potassium soap solutions show the same cleaning action? If not, explain.

b. Describe how the soap molecules functioned to remove the grease spot.

c. Did your potassium soap solution form a precipitate with any of the metal ions that were added? If so, which?

d. Write equations for any reactions that occurred between the soap solution (use a typical fatty acid residue) and the metal ions.

e. How do sodium, potassium, and calcium soaps compare in solubility?

f. What ions are responsible for the qualities of "hard water"?

g. Did the metal ions form precipitates with the detergent molecules? If not, why not?

h. Is the tap water in your laboratory "hard"? What is your evidence?

Proteins and Amino Acids | 30

OBJECTIVES

1. To learn how the chemistry of acids and amines applies to amino acids and proteins.
2. To become familiar with some chemical tests that distinguish between amino acids and proteins.
3. To compare the properties of a class of natural polymers (the polyamides we call proteins) with those of the monomers (amino acids).
4. To examine some common food products that contain proteins and amino acids.

DISCUSSION

Proteins are the stuff of life. The heart of every cell of every living organism is protein. Proteins are natural polymers, which perform a host of important functions in living systems. They carry oxygen, provide a unique shell (skin), catalyze reactions, control pH, transmit nerve impulses, and lift objects (muscles). Protein in the diet supplies food energy and amino acids. Amino acids are the monomers in proteins.

In today's experiment you will examine the properties of amino acids and proteins, observe some of their reactions, and conduct tests to distinguish among them.

Amino Acids

Let us first consider some properties of the monomers and then examine how they bond together to form polymers (proteins). Most amino acids exhibit one common feature: they have a primary amino group on the α-carbon, the carbon next to the carbonyl group.[1]

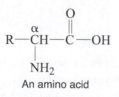

An amino acid

The twenty-odd amino acids found in proteins have different R–groups. Examples of three amino acids that you will study in today's experiment are alanine, tyrosine, and glutamic acid. Others that are important as a background for your experiment are glycine, lysine, tryptophan, cysteine, and cystine (table 30.1).

---

[1]The structures of amino acids are somewhat more complicated than has been indicated here. Amino acids exist as dipolar ions (called *zwitterions*) in the solid state and also to a large extent in aqueous solution.

$$R-\underset{\underset{NH_3^+}{|}}{CH}-\overset{\overset{O}{||}}{C}-O^-$$

TABLE 30.1 Characteristics of Some Amino Acids

Amino Acid	R–Group	Comment	Essential*
Glycine	H—	Simplest	No
Alanine	CH_3—	Aliphatic	No
Tyrosine	HO—⟨⟩—CH_2—	Aromatic	No
Glutamic acid	HOOC—CH_2CH_2—	Acidic	No
Lysine	$NH_2CH_2(CH_2)_2CH_2$—	Basic	Yes
Cysteine	HS—CH_2—	Contains sulfur	No
Cystine	—CH_2SSCH_2—*	Disulfide, oxidation product of cysteine	No
Tryptophan		Heterocyclic and aromatic	Yes

*Essential amino acids cannot be synthesized in the body and must be obtained from protein in the diet.

The Structure of Proteins

Proteins are huge molecules with molecular weights of 10,000 to more than 1 million. They are composed of amino acid monomer units, called residues, which are bonded together by peptide (amide) bonds. The precise order of the amino acid residues is known as the primary structure of the protein.

Proteins have precise, three-dimensional configurations. They frequently contain α-helix and pleated sheet structures, called secondary structures. Tertiary structure refers to the folding and coiling of a peptide chain (already arranged in a particular secondary structure) to produce the complex, rather rigid overall conformation. Hydrogen bonds, disulfide bonds (cystine), salt bridges, and other types of bonds maintain the protein in its distinct shape. Try to picture in your mind this complex, coiled structure as you carry out your experiments.

Denaturation of Proteins

Destruction of the three-dimensional shapes of protein molecules by various physical and chemical means is called *denaturation*. Proteins become denatured by any action that ruptures hydrogen bonds or disrupts salt bridges. Such action includes heating; mechanical stress (such as whipping cream or bruising skin), and treatment with ultraviolet light, organic solvents (alcohols), acids, bases, and detergents. Heavy metal ions, such as mercury, silver, and lead, denature primarily by disruption of salt bridges. You will employ several of these methods of denaturation in today's experiment.

Hydrolysis of Proteins

As you would expect from knowing that they are polyamides, proteins can be hydrolyzed in acidic or basic solution to produce free amino acids. The reaction is illustrated with a tripeptide in which the amino acid residues are enclosed and the amide bonds are indicated by arrows. The hydrolysis process is so slow, however, that in today's experiment a previously hydrolyzed protein will be provided for you.

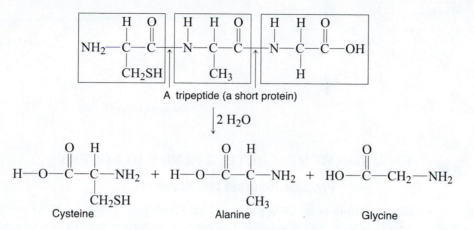

A tripeptide (a short protein)

Cysteine   Alanine   Glycine

The following tests are used to detect the presence of amino acids and proteins and to distinguish between them.

The Biuret Test

The biuret test is positive for proteins but not for amino acids. The evidence for the test consists of the formation of a violet-pink complex when cupric ion, Cu^{2+}, in basic solution is added to any polymer such as protein, which contains multiple amide bonds.

The Xanthoproteic Test

This test depends upon a reaction with a specific type of amino acid side chain. The xanthoproteic test is positive for the side chains in tyrosine and tryptophan. Since these amino acid residues (especially tyrosine) are very common in proteins, the test is positive for most proteins. The reaction occurs with the two free amino acids as well. The xanthoproteic test involves replacing some of the hydrogens of the reactive aromatic rings with nitro groups (nitration with concentrated nitric acid), followed by reaction with base to produce an orange-yellow color.

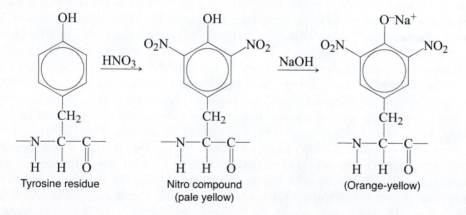

Tyrosine residue   Nitro compound   (Orange-yellow)
                     (pale yellow)

The Ninhydrin Test

The ninhydrin test is positive for amino acids and some proteins. The test involves several complex reactions but can be summarized as follows: Two ninhydrin molecules remove the alpha amino ($—NH_2$) group from an amino acid or protein to form a blue-violet complex and other compounds.

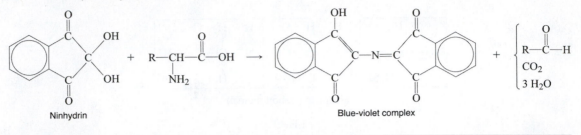

Ninhydrin   Blue-violet complex

ENVIRONMENT, CULTURE, AND CHEMISTRY
Proteins: Sources and Problems

Proteins in our diet come from meat and vegetables. Meat (beef, pork, poultry, or fish) has considerably more protein (approximately one third) than black or kidney beans, the richest source of vegetable protein; cheese products, milk, and eggs have less protein than meat but more than vegetables. The amount of protein in cooked vegetables, such as broccoli and green beans, is about 75% less than in black or kidney beans. Even related products do not have the same amount of protein: duck has more than turkey and lentil beans have more than navy beans. One problem with vegetable protein is that not all the proteins from these individual vegetable sources meet the essential amino acid requirement; nine, but most books say eight essential amino acids cannot be synthesized (made) by human cells and, therefore, we depend on them being supplied in our diet. Vegetarians (vegans) can meet the essential amino acid requirement by eating a combination of different grains, legumes (for example, beans, peas, soy products), nuts, and seeds. There are distinct advantages in vegetable protein: fiber, vitamins and minerals, less fat, and no cholesterol. Red meat, while a great source of protein, contains a high percentage of saturated fat, which is probably connected to cardiovascular disease. Fish has about 30% less fat than red meat, and most of it is unsaturated, "good" fat. Another serious problem with meat is its source; today, most meat comes from "animal factories." When we think of cows, pigs, and chickens, our meat sources, our minds and memories are filled with the images of bucolic, green pastures. These images, however, are far from the truth of the modern "animal factory" where animals are crowded together in pens, fed rich foods so they will grow rapidly, and not allowed to dissipate the weight with exercise. The animals are in close association with their own filth and under constant stress, which weakens the immune system. Disease would be rampant under these unhealthy circumstances if it were not for high doses of antibiotics. This use of antibiotics, almost certainly overuse (70% of all antibiotics in the US are fed to farm animals), results in bacteria that become drug resistant. These antibiotic-resistant bacteria are then passed on to humans. In addition, the animal waste (manure), which is spread on the fields as fertilizer, is a continual source of antibiotics in the environment; it is estimated that 90% of the antibiotics administered to animals is not metabolized. Rain and irrigation water carry the antibiotics to human water sources, where they can enter the body and lead to drug resistance. An equally important issue is the treatment of the animals. They are living creatures whose lives will eventually be sacrificed for our benefit. Should they be treated with greater care? Meat that is produced more slowly in green pastures with trees for shade on a hot summer day will cost more, but perhaps that is the price that we should pay. Fortunately, society has begun to pass laws requiring greater living space for animals while they are caged and less painful deaths in slaughterhouses. Humans need to ponder the fact that they share this world with all living creatures and what their responsibilities are to them.

Pre-Laboratory Questions | 30

1. Define the terms *dialysis, hydrolysis, residue, denaturation, and polypeptide.*

2. Explain why glutamic acid is an acidic amino acid and lysine is a basic amino acid.

3. Would a tripeptide give a positive biuret test? Explain.

4. Which of the following tests can distinguish between a protein and an amino acid: biuret, ninhydrin, and xanthoproteic? Give the reason for your choice(s).

EXPERIMENT

1. Solutions of Amino Acids and Proteins

Egg white solution. Break an egg into a 400-mL beaker, separating the white of the egg from the yolk and discarding the yolk. Add 300 mL of water to the white of the egg, and stir vigorously for a short time. Filter the liquid through a piece of cheesecloth into a clean beaker. Divide the egg albumin solution between four students.

   Gelatin solution. Obtain 20 mL of a 1% gelatin solution.

   Amino acid solutions. Obtain 10 mL of a 1% alanine solution, 10 mL of a 1% glutamic acid solution, and 10 mL of a saturated solution of tyrosine.

   Note the appearance of the protein dispersions and the amino acid solutions. Keep all of the solutions for experiments that follow.
(Answer Question 1 on the Observation and Results sheet.)

2. Observations of a Protein Hydrolyzate (Product of Protein Hydrolysis)

Since the hydrolysis of proteins requires several hours of heating with strong acid, you will make observations on a protein hydrolyzate. One readily available source of protein hydrolyzates is bouillon. Obtain a bouillon cube (chicken bouillon is recommended, since it gives a lighter-colored solution), and write down the list of ingredients from the label. Prepare a solution of the bouillon cube by dissolving 1 g of it in 50 mL of warm water. Share the solution with another student. This solution will be used for some later tests. (Answer Question 2, a–e.)

3. Dialysis of Proteins and Amino Acids

In this experiment, you will work in pairs to compare the abilities of proteins and amino acids to pass through (dialyze) semipermeable membranes. All should participate in setting up the apparatus and in selecting the concluding tests.

   Obtain three 30-cm pieces of dialysis tubing (2.5 cm in diameter). Tie a string tightly around the center of each piece of tubing (figure 30.1). Wet the tubing so that it can be opened more easily, and open one end of the tubing with a stirring rod so as to make a sack. Pour 10 mL of egg albumin solution, tyrosine solution, and bouillon solution into different sacks. Lift up the empty bottom half of each piece of tubing and tie it and the top of the sack tightly shut. *Wash off any of the solution that may have spilled on the outside of the sacks with distilled water*. Place the sacks in large (2.5 × 15 cm) test tubes, and add 15 mL of distilled water to the test tubes. (The tops of the sacks must remain above the water line or leakage may occur.) Allow the solutions to stand for about an hour at room temperature. Occasionally swirl the solutions in the test tubes during this time. After an hour or so, perform tests of your own choosing to determine whether diffusion into the surrounding solution has occurred. (Answer Question 3, a–d.)

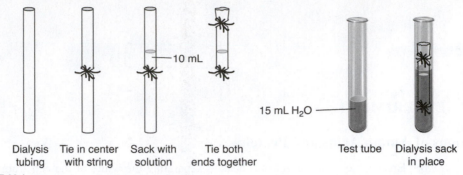

▲ **FIGURE 30.1** Preparation of a dialysis sack.

4. The Biuret Test

Make tests on solutions of gelatin, glutamic acid, egg albumin, and bouillon as follows. Mix 2 mL of the solution to be tested with 2 mL of 10% NaOH, and then add exactly 2 drops of 0.5% cupric sulfate. Stir the solution thoroughly. A positive test is formation of a pink or violet color. (Answer Question 4, a–e.)

5. The Ninhydrin Test

Compare results for solutions of egg albumin, bouillon, and alanine, using the following procedure. Mix 1 mL of 0.2% ninhydrin solution and 2 mL of the solution to be tested in a test tube and heat for several minutes in a boiling water bath. Observe any color formation. (Answer Question 5a, 1–3; and 5b.)

6. The Xanthroproteic Test

Make tests on solutions of tyrosine, alanine, and egg albumin according to the following procedure. Add 10 drops of concentrated nitric acid to 1 mL of the solution to be tested and heat for several minutes in a boiling water bath. Cool the test tubes in cold water and then (cautiously—much heat will be released) add 10% sodium hydroxide, stirring constantly, until the solution is alkaline to litmus. A positive test is an orange-yellow color. Assemble the test tubes in order of increasing intensity of color and record the order. (Answer Question 6, a–e.)

> **Caution**: Be careful not to get nitric acid on your skin, which is also made of protein.

7. Denaturation of Proteins

Heat 1 mL of the egg albumin solution and 1 mL of one of the amino acid solutions in separate test tubes directly over a flame. Watch for the formation of a gelatinous solid on the surface of the solution. Remove any solid with a wood splint and examine it. (Answer Question 7, a–d.)

Mix 2 mL of 95% ethyl alcohol with 1 mL of egg white solution. Is there any evidence of coagulation? (Answer Question 7e.)

8. Effect of Heavy Metals on Protein and Amino Acid Solutions

Mix several drops of 1% lead nitrate solution with 1-mL portions of your egg albumin, gelatin, and glutamic acid solutions in separate test tubes. (Answer Question 8, a–d.)

> **Dispose of all waste from each test in the appropriate containers.**

QUESTIONS AND PROBLEMS

1. Sketch α-helix and pleated sheet structures. Define the relationship of these structures to secondary and tertiary structures.

2. Show the primary structure of the polypeptide whose abbreviated structure is Gly-Phe-Trp-Val-His. (Assume that the free amino group is on the residue on the extreme left.) Write an equation for the hydrolysis of this polypeptide in acidic solution.

3. What happens to the three-dimensional structure of a protein during whipping of an egg to make meringue or during cooking of an egg? Indicate which bonds are broken.

4. When concentrated nitric acid comes into contact with the skin, a yellow coloration occurs. On the basis of what you learned in this experiment, suggest an explanation for this observation.

5. One of the symptoms of nephritis is the appearance of albumin in the urine. Suggest a method whereby you could detect the presence of albumin.

6. Alcohol is often used to sterilize medical instruments or a patch of the skin. On the basis of this experiment, indicate the function of alcohol in these operations.

7. Corn protein (zein) is low in lysine and tryptophan. Why is it an inadequate (or incomplete) protein?

8. Suppose that a protein were composed entirely of glycine, alanine, leucine, and valine. Draw the structure of a segment of this protein, using all of the amino acid residues. Would this protein be an effective buffer? Explain.

9. This experiment did not include the isolation of a pure amino acid from the hydrolysis product of a protein such as bouillon. Why would the isolation of a pure amino acid be extremely difficult?

EXPERIMENT

Proteins and Amino Acids | 30

OBSERVATIONS AND RESULTS

1. Solutions of Proteins and Amino Acids

Describe the difference in appearance between the protein and amino acid solutions, and explain why this difference exists.

2. Observation of a Protein Hydrolyzate (Product of Protein Hydrolysis)

a. List the ingredients in a bouillon cube.

b. Indicate which ingredients refer to amino acids and which to proteins.

c. Suggest how the protein might have been hydrolyzed to give the amino acids.

d. It takes much longer for amino acids from ingested protein to appear in the bloodstream than it does for those from bouillon. Explain.

e. Under what conditions would bouillon be preferred to meat in the diet?

3. Dialysis of Proteins and Amino Acids

a. List the three solutions and the corresponding tests that you propose to use to determine whether dialysis has occurred.

b. Now list the result of each test and your conclusions concerning dialysis.

c. What do you conclude concerning the relative abilities of protein and amino acid molecules to pass through (permeate) the dialysis tubing?

d. Account for this difference between proteins and amino acids.

4. The Biuret Test

Record the color for each solution and indicate whether it confirms a positive or a negative test.

a. Gelatin:

b. Glutamic acid:

c. Egg albumin:

d. Bouillon cube:

e. What chemical structure is required for a positive biuret test? Draw this structure.

5. **The Ninhydrin Test**

   a. Indicate whether the solutions gave a positive or a negative test and record the color:

   1. Egg albumin:

   2. Bouillon:

   3. Alanine:

   b. What functional group is required for a positive test? Do both proteins and amino acids have this group? Explain.

6. **The Xanthoproteic Test**

   a. Rank your solutions in order of increasing color intensity.

   b. Which solutions(s) gave a negative test? Explain why the solutions(s) gave a negative test.

   c. Which solution gave the most intense color? Give an explanation.

   d. What chemical reaction occurred in all the solutions that gave positive tests? Use structures to show the reaction.

   e. Does this test confirm the presence of specific amino acid residues in the proteins? Explain.

7. Denaturation of Proteins

a. Describe any precipitate that formed when the protein and amino acid solutions were heated.

b. If a change occurred in one solution but not the other, give an explanation for this.

c. Explain your observations from the standpoint of change in the three-dimensional structure of the protein.

d. What happened to the primary structure of the proteins during heating and treatment with alcohol?

e. Was there evidence of coagulation when the egg white solution and ethyl alcohol were mixed? Describe the evidence.

8. Effect of Heavy Metals on Protein and Amino Acids Solutions

a. Describe what happened when you added lead nitrate to the three solutions.

b. What structure in a protein is required for a reaction with lead nitrate? Draw this structure.

c. Why was the lead nitrate positive for egg albumin but not gelatin?

d. Describe how your observations could account for (1) the generally toxic effects of heavy metals and (2) the use of egg white as an antidote for accidental ingestion of heavy metals.

Enzymes and Chemical Reactions | 31

OBJECTIVES

1. To prepare solutions of some enzymes.
2. To become familiar with some of the physical properties of enzymes.
3. To observe specific reactions that are catalyzed by enzymes.
4. To learn to recognize factors that affect the catalytic activity of an enzyme.

DISCUSSION

Of the thousands of reactions occurring in a living organism, all but a few are catalyzed by specific enzymes. The remarkable thing about enzymes is their ability to enable reactions that normally occur slowly or not at all at body temperatures to take place at greatly increased speeds, often so fast that they seem instantaneous.

In today's experiment you will prepare solutions of some enzymes and observe their effect on the rates of reactions. You will also examine factors that increase and decrease the effectiveness of enzymes.

Classification and Terminology

Enzymes are named and classified either according to the reaction that they catalyze or after the substrate (reactant) upon which they act. Hence, they are called hydrolases (hydrolysis), oxidases (oxidation), ureases (the substrate is urea), and so forth. For enzymes to function, non-protein cofactors called coenzymes (organic molecules) or metal activators (metal cations) are usually required. Technically, the protein part of the total enzyme system is called an apoenzyme. In this experiment, we will use the term *enzyme* for *apoenzyme*.

Properties of Enzymes

Enzymes exhibit all of the general properties of globular proteins. Most enzymes are soluble in water and can be hydrolyzed to amino acids. Enzymes are easily denatured (see Experiment 30) by extremes of temperature and a variety of chemical reagents, with concomitant loss of enzymatic activity. You will recall that denaturation of a protein involves destruction of its three-dimensional shape.

How Enzymes Speed Reactions

Enzymes react with substrates to form activated complexes. Functional groups on the substrate and on the side chains (R-groups) of the amino acid residues in the enzyme form weak bonds that hold the complex together. In the activated complex, substrate bonds are broken and new ones are made, to give the product and then re-form the enzyme.

The small portion of the surface of an enzyme to which the substrate bonds is called the active site. The active site may be destroyed by denaturation (the R–groups involved in bonding at the active site are moved out of position). It may also be blocked by combination with a non-substrate molecule, called an inhibitor. Inhibition results in partial, and in some cases total, loss of enzyme activity. These concepts are shown diagrammatically in figure 31.1.

Two Enzymes: α-Amylase and Catalase

In today's experiment you will observe the catalytic activity of two different enzymes. The first of these is salivary amylase. This enzyme is found in saliva and is classified as an α-amylase because it catalyzes the hydrolysis of glycoside (acetal) bonds in starch at random points (review the hydrolysis of starch in Experiment 26). The result is rapid conversion to small starch molecules (dextrins), which do not give the blue-black complex with iodine. In time, the dextrins are converted to a mixture of maltose and glucose.

$$\text{Starch} \xrightarrow[\text{rapidly}]{\alpha\text{-amylase}} \text{dextrins} \xrightarrow[\text{slowly}]{\alpha\text{-amylase}} \text{maltose} + \text{glucose}$$

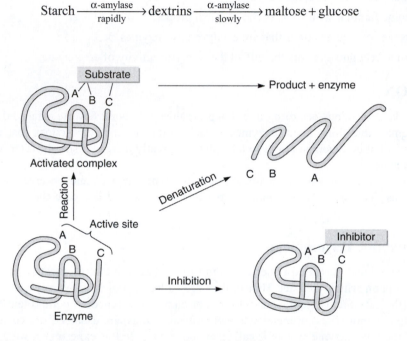

▲ **FIGURE 31.1** Diagram of enzyme reaction, denaturation, and inhibition.

The second enzyme, catalase, catalyzes the decomposition of hydrogen peroxide, H_2O_2.

$$2\,H_2O_2 \xrightarrow{\text{catalase}} 2\,H_2O + O_2$$

Catalase is an enzyme that contains iron-heme. Cyanide ion (CN^-) strongly inhibits this enzyme through formation of a complex with the iron ion (Fe^{2+}) in the heme ring. Although catalase occurs widely in nature, particularly in plants, its role in the oxidative scheme is not completely understood. The crude extract of potato, from which you will obtain catalase, contains many enzymes in addition to catalase.

ENVIRONMENT, CULTURE, AND CHEMISTRY
Enzymes: They Can Be Both Good and Bad

Enzymes have many connections to our economy and culture: some good and some bad. A large number (20–30%) of adults in the United States cannot digest dairy products because they lack the enzyme lactase, which is required to catalyze the hydrolysis of lactose, a carbohydrate (sugar) in milk. As a result, when they ingest dairy products, they develop distressing symptoms such as hives, stomach pain, vomiting, diarrhea, constipation, and flatulence. Fortunately, relief is provided by taking a pill containing the enzyme itself or the dairy product can be pretreated with the enzyme. Pompe disease is a less known, but serious, illness resulting from the body's inability to make the enzyme lysosomal acid alpha-glucosidase, called acid maltase. Both children and adults can exhibit this disease. The average age of death in children is about nine months. The symptoms in children are a floppy appearance, delayed motor skills development, and feeding difficulties. Adults show a persistent cough, difficulty in swallowing or chewing, and progressive muscle weakness. Again, this disease is successfully treated by providing the enzyme as a medical treatment. Currently, a corporation is building a multimillion dollar plant to mass produce this enzyme. Yet another disease, Tay-Sachs disease, appears primarily in babies around six months of age and is caused by the lack of the enzyme hexosaminidase; absence of the enzyme allows gangliosides, a fatty material, to build up in the brain. Babies with Tay-Sachs gradually stop smiling and crawling, become blind and paralyzed, and usually die by age five. The disease is found almost entirely in babies of Jewish heritage. In order for the baby to inherit Tay-Sachs, both parents must be genetic carriers, and then there is a 25% chance of inheriting the disease. Finally, to round out the bad side of enzymes, the illnesses and deaths from snake bites (a large number each year worldwide) are caused primarily by enzymes in the venom.

Now for the good side of enzymes: enzymes in the blood can be useful as indicators, providing an early warning of a particular disease while there is still time for treatment. High levels of creatine phosphokinase, aminotransferase, and an enzyme BACEl may indicate a recent heart attack, disease of the liver, and the onset of Alzheimer's Disease, respectively. Other enzymes are finding valuable uses in our society as catalysts. This is important because enzymes are readily degraded in nature to amino acids that are food for many microorganisms. Proteases, a group of enzymes, have become valuable in laundry and dishwasher detergents as cleaning agents, replacing phosphate, which caused environmental problems. The fabric in jeans can be given "the used look" with the enzyme cellulase; previously the jeans were stone-washed, which damaged the fabric. Catalase, the enzyme that is the center of today's experiment, has found an industrial use in the fabric industry by oxidizing the hydrogen peroxide in yarn production prior to dyeing. The cheese industry has long depended on rennet, an enzyme extract from the stomachs of slaughtered calves. The gene for rennet has now been transferred to a bacterium that produces the enzyme chymosin, which is purer and more effective in cheese production. Thankfully, calf stomachs are no longer needed. Some of the highest quality baked bread once required the toxic chemical bromate in the treatment of flour. Now, an enzyme, glucose oxidase, is used as a substitute for bromate. A current goal of chemists is to create an artificial enzyme (completely protein), which will speed up important reactions and negate the need for conventional catalysts such as metals and strong acids and bases, which contaminate the environment. Recently an instance of that goal was attained. An artificial enzyme called chemzyme was synthesized, which was capable of breaking down the food-borne toxin glycoside esculin.

Pre-Laboratory Questions | 31

1. Read the discussion section and the experimental procedure of Experiment 30 as review for today's experiment.

2. Define, in your own words, the terms *amino acid residue, active site, catalysis, denaturation, reaction rate increase, and enzyme specificity.*

3. In today's experiment, what is the substrate in the reaction that is catalyzed by the enzyme catalase?

4. What are the monomers from which α-amylase is synthesized?

5. Draw the structures of glucose and maltose and segments of the structures of amylose and amylopectin.

EXPERIMENT

1. α-Amylase

Collect about 5 drops of saliva, containing α-amylase, in a small clean beaker. (Stimulate the secretion of saliva by chewing a small piece of paraffin wax.)

Fill a 400-mL beaker two-thirds full of water and heat the water to approximately 35°C; this temperature must be maintained within a few degrees for optimal enzyme activity. Now make a saliva solution by mixing 2 drops of saliva with 2 mL of distilled water. Add the saliva solution to 4 mL of 1% starch solution and stir. Familiarize yourself with the starch-iodine test by *immediately* adding 1 drop of iodine solution to a few drops of the starch-saliva solution on a watch glass. Note the color.

Place the test tube containing the rest of the α-amylase (saliva) and starch in the 35°C water bath. Every 2 or 3 minutes remove a few drops of solution and perform the starch-iodine test. When the starch-iodine test is negative (record this time), confirm the presence of reducing sugar by using Fehling's test. (See Experiments 21 and 26 for a review of Fehling's test.) (Answer Question 1, a–f on the Observations and Results sheet.)

Dispose of waste in the appropriate container.

2. Catalase

Work in pairs. Prepare a potato extract by grinding 25 g of peeled, finely sliced potato in a large, clean mortar or an evaporating dish. When the potato has become a fine pulp (after about 5 minutes), add 100 mL of distilled water and continue to grind and stir the pulp for several minutes more. Filter the pulp through cheesecloth stretched over a beaker or through glass wool packed into a funnel (figure 31.2).

In this experiment you will make some quantitative observations on the rate of decomposition of hydrogen peroxide catalyzed by catalase. Set up the apparatus shown in figure 31.3. Fit a 25-mL Erlenmeyer flask with a rubber stopper containing a piece of right-angle glass tubing attached to a piece of rubber tubing 30 cm long. Insert a U-shaped glass tube into the other end of the rubber tubing.

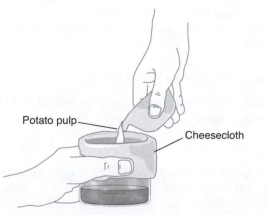

Potato pulp

Cheesecloth

▲ **FIGURE 31.2** Filtering potato pulp through cheesecloth.

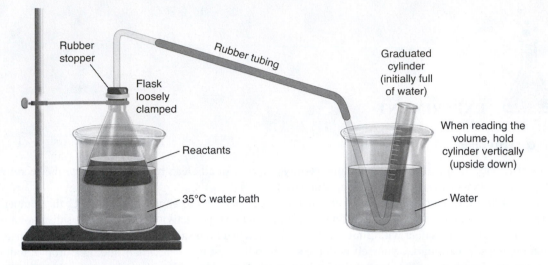

▲ **FIGURE 31.3** Apparatus for the reaction of catalase with hydrogen peroxide. The volume of oxygen collected indicates the extent of the reaction.

Caution: *Be sure that all connections are gas-tight.* To measure the oxygen that evolves, invert a 10-mL graduated cylinder (filled with water) into a 250-mL beaker that is two-thirds full of water.

Before proceeding to the quantitative parts of the experiment, carry out a trial run so that you will know how to manipulate the equipment. Have the graduated cylinder filled with water and inverted in the beaker of water. To the 25-mL Erlenmeyer flask add 5 mL of catalase solution and 15 drops of 3% hydrogen peroxide. Attach the gas delivery tube and place the flask in the 35°C water bath. Swirl the solution occasionally and wait for about 30 seconds for the temperature to equilibrate. Then place the end of the delivery tube in the graduated cylinder. Note that the gas is displacing the water from the cylinder. Observe the reaction (answer Question 2, a–c). If you are not familiar with the reaction, proceed to the quantitative studies; otherwise repeat the procedure.

A. Temperature of 35°C: The General Procedure

Mix 5 mL of catalase solution (potato extract) with 15 drops of 3% hydrogen peroxide in the 25-mL flask, swirl the flask to mix, and place in the water bath at 35°C. Wait about 30 seconds, and collect oxygen gas for 2 minutes. Swirl the solution occasionally during this time. The volume of oxygen is, of course, the volume of displaced water at the "top" of the inverted graduated cylinder, subtracting the volume of any air pocket that may have been present before the oxygen was collected. Rinse the reaction flask with distilled water after each run. *Since this was your first run, repeat the determination and average the results.* **For Experiments A–G, place all volume of oxygen data beside the specific reaction condition in the table of 2, d.**

B. Lower Concentration of Hydrogen Peroxide

Proceed as in 2A, except add 5 drops of hydrogen peroxide instead of 15 drops. Record the volume of oxygen produced in 2 minutes. (Answer Question 3, a–b.)

C. Temperature of 0°C

Proceed as in 2a, except, *before* adding the 15 drops of hydrogen peroxide, swirl the 5 mL of catalase solution in a mixture of ice and water until the temperature drops to approximately 2°C. Keep the flask in the ice bath during the time the oxygen is collected. Record the volume of oxygen produced in 2 minutes. (Answer Question 3c.)

D. Enzyme Boiled Before Use

Proceed as in 2a, except gently boil (*using your burner*) 5 mL of catalase solution in the reaction flask for 15 seconds. (Note any changes in appearance of the enzyme solution.) Cool the flask immediately in the ice-water bath, and then bring it to 35°C in the water bath. Next, add the hydrogen peroxide and carry out the reaction in the 35°C water bath. Record the volume of oxygen produced in 2 minutes. (Answer Question 3, d–e.)

E. pH Change

Proceed as in 2a, except place 5 mL of catalase solution in the flask. Use broad-range pH paper to determine the pH of the solution. Now add 5 drops of 10% sodium hydroxide solution to the catalase solution, and again determine the pH. (Record the initial and final pH of the solution.) Add 15 drops of hydrogen peroxide, and place the flask in the 35°C water bath. Record the volume of oxygen produced in 2 minutes. (Answer Question 3, f–g.)

F. Change in Structure of the Substrate

Proceed as in 2a, except use 15 drops of a 3% solution of *tert*-butyl hydroperoxide, $(CH_3)_3 C—O—O—H$, in place of hydrogen peroxide. Record the volume of oxygen produced in 2 minutes. (Answer Question 3, h–i.)

G. Addition of Cyanide

DEMONSTRATION: Because of the danger of working with potassium cyanide, your instructor may demonstrate this part of the experiment using a solution of potato extract from your class. Proceed as in 2a, except mix 5 drops of 0.05 M potassium cyanide, KCN, with 5 mL of potato extract and shake the mixture. Add the hydrogen peroxide and place the flask in the 35°C water bath. Record the volume of oxygen produced in 2 minutes. (Answer Question 3j.)

Caution: *Cyanides are extremely poisonous. Wash your hands after using cyanides and dispose of all cyanide solutions as directed by your instructor. Rinse your flask with distilled water.*

Dispose of all waste from procedures 2, A–G in appropriate containers.

QUESTIONS AND PROBLEMS

1. In the body, enzymes are located primarily in the intracellular fluids. On the basis of today's experiment, explain why it is essential that these fluids be strictly buffered.

2. The decomposition of carbonic acid, H_2CO_3, to carbon dioxide and water is an extremely fast reaction. Even so, during the short time that the blood spends in the capillaries of the lung the uncatalyzed reaction does not occur fast enough to allow CO_2 to escape from the blood into the lung. An enzyme, carbonic anhydrase, increases the rate of this reaction 600 times. Write an equation for this reaction. Draw the structure of H_2CO_3, showing all bonds, and indicate which bonds are broken and made in the active site. What is the source of the carbonic acid in the blood?

3. An enzyme called β-amylase, which is found in malt, hydrolyzes starch (amylose) to maltose. Draw a segment of a starch molecule and indicate the points of attack by α-amylase and β-amylase.

Enzymes and Chemical Reactions	EXPERIMENT 31

OBSERVATIONS AND RESULTS

1. α-Amylase

a. How much time elapsed before the starch test was negative?

b. What products (give names) are formed in this hydrolysis reaction?

c. Was Fehling's test for a reducing sugar positive?

d. Use a starch molecule (a segment) and a water molecule to show the reaction that is catalyzed by an α-amylase.

e. On the basis of today's experiment, discuss the function of α-amylase in the saliva.

f. Suggest several eating habits that would decrease the effectiveness of α-amylase.

2. Catalase

a. Describe the reactions that occurred when you added the hydrogen peroxide to the potato extract.

b. Write an equation for the reactions.

c. Catalase exerts its effects on the hydrogen peroxide molecule by breaking the oxygen-to-oxygen (peroxide) bond. Write the structure of hydrogen peroxide and indicate with an arrow the bond that is broken.

d. Complete the following table showing the volumes of oxygen produced in the reactions of catalase with hydrogen peroxide under various conditions.

Experiment	Specific Reaction Conditions	Volume of Oxygen (after 2 minutes)
1.	the general procedure (standard)	
2.		
3.		
4.		
5.		
6.		
7.		

3. a. Express, in *decimal form*, the ratio $(vol_{2.d2}/vol_{2.d1})$ of the rates (volumes of oxygen produced) under dilute (b) and concentrated (a) conditions. Then express, in decimal form, the ratio $(drops_B/drops_A)$ of concentrations (number of drops of H_2O_2 solution under these same conditions). From the standpoint of these two ratios, is there a relationship between the change in concentration and the change in rate? Explain.

b. Account for the difference in rates for the two concentrations on the basis of frequency of collisions of enzyme and substrate molecules.

c. Give an explanation for the difference in rates that was observed at 0°C and 35°C.

d. After the catalase solution was heated to 100°C was the rate of reaction slower or faster? Explain the change in rate.

e. Did the appearance of the enzyme solution change when it was heated? If so, how did it change? What aspect or property of the enzyme was responsible for change?

f. For part 2E what was the initial pH? What was the final pH? How was the rate of the reaction affected by the increase in pH?

g. Did this change of pH in part 2E affect the configuration of the active site? What amino acid residue in catalase would be changed by a more basic pH? Show the reaction of a residue with sodium hydroxide.

h. Draw the complete structure of *tert*-butyl hydroperoxide and hydrogen peroxide, showing all bonds, and circle the peroxide bond.

i. Did *tert*-butyl hydroperoxide fit into the active site in catalase as well as hydrogen peroxide? What is your evidence?

j. What effect did cyanide ion have on the rate of the reaction? What term is used to describe the effect of cyanide? Was the active site affected by cyanide ion? Explain.

Analysis of an Important Food: The Peanut | 32

OBJECTIVES

1. To detect the presence of several of the biochemical constituents in a food product by application of chemical tests.
2. To isolate and determine the percentage of an important constituent of the peanut—peanut oil.

DISCUSSION

The peanut is an important, balanced food placed on the map years ago by George Washington Carver and more recently by former President Carter. Peanuts contain more protein, more minerals, and more vitamins than an equal weight of beef liver. Nutritionists list peanut butter as a meat substitute. Peanuts are rich in fat, containing more fat per volume than heavy cream. Half of the peanuts harvested in the United States are ground into peanut butter. Most of the remainder are sold as roasted salted nuts or are used to make candy and baked goods. In today's experiment you will identify several of the biochemical constituents of the kernel and shell of the peanut and isolate an important lipid, peanut oil.

Chemical Composition of the Kernel

The kernel of the peanut is reported to have the following composition: water, 3%; lipid, 50%; protein, 25.8%; carbohydrate, 18.7%; and inorganic salts, 2.5%. Calcium and iron are among the salts. Significant amounts of the vitamins thiamine, riboflavin, and niacin are present. The percentages of the fatty acid residues in the triglycerides of the lipid peanut oil are: palmitic, 6% to 9%; stearic, 2% to 6%; oleic, 50% to 70%; and linoleic, 13% to 26%. Its iodine number is between 83 and 98.

Isolation of Peanut Oil

Separation of peanut oil from the peanut is based on the fact that fats (triglycerides) are soluble in organic solvents, such as methylene chloride, CH_2Cl_2, whereas the other biochemical constituents (carbohydrate and protein) are not. Therefore, you will isolate peanut oil by extracting ground peanuts with methylene chloride, separating the solvent from the meal, and evaporating the solvent.

Carbohydrates in the Shell of the Peanut

The shell of the peanut contains at least two carbohydrates. One of these polysaccharides is cellulose, which you studied in Experiment 27. The other polysaccharide, xylan, is composed of the monomer (monosaccharide) xylose. Structures of xylose and a segment of xylan are shown.

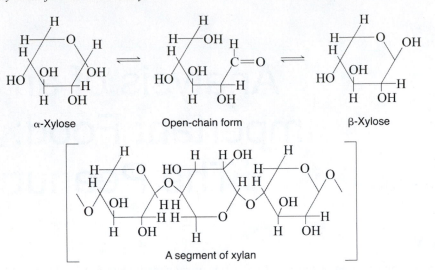

Tests for the Biochemical Constituents

You will use Fehling's test (see Experiments 21 and 26), the xanthoproteic test (see Experiment 30), and precipitation with ammonium oxalate (see Experiment 16) to detect carbohydrates, protein, and calcium, respectively.

Potassium permanganate, $KMnO_4$, solution will be used to test for unsaturated fatty acid residues in peanut oil. You will obtain evidence that glycerol is the "alcohol portion" of the triglyceride with the acrolein test. This test consists of heating a triglyceride with potassium acid sulfate, $KHSO_4$, to form glycerol, as well as a variety of other products. The glycerol reacts further with $KHSO_4$ to give acrolein, an unsaturated aldehyde with a sharp, penetrating odor.

$$\text{A triglyceride} \xrightarrow{KHSO_4} \underset{\substack{| \quad | \quad | \\ OH \quad OH \quad OH \\ \text{Glycerol}}}{CH_2-CH-CH_2} \text{(and other products)} \xrightarrow{KHSO_4} \underset{\text{Acrolein}}{CH_2{=}CH-\overset{\overset{\displaystyle O}{\|}}{C}-H}$$

Pentoses (five-carbon monosaccharides) such as xylose can be detected with Tollens' phloroglucinol test. This test consists of the reaction of the pentose and phloroglucinol to give a reddish product.

ENVIRONMENT, CULTURE, AND CHEMISTRY
Peanuts and Other Allergies

Only a few years ago, the first snack that one was given at the beginning of a plane flight was a small bag of peanuts. Not anymore. What happened to stop this almost universal introduction to a plane trip: peanut allergy? A small number (around 1%) of people, particularly children, are allergic to peanuts, sometimes with catastrophic results. In fact, peanut allergies are the most common cause of food-related deaths. What is responsible for these often severe reactions? Ingestion of peanuts causes the release of enzymes (antibodies), which in turn stimulate cells (mast cells) to release various chemicals, typically histamine, into the bloodstream. The presence of histamine, an organic amine, results in bronchospasms, a constriction of the airways in the lungs. Other symptoms may include one or more of the following: vomiting, diarrhea, hives, swelling of the lips, face, throat and skin, severe abdominal pain, and an asthma attack. The most severe reaction is anaphylactic shock, which requires immediate life-saving administration of a counter-active drug such as epinephrine or Benadryl. Fortunately, about 25 percent of children outgrow the allergy to peanuts. There is no known cure for peanut allergy, so persons afflicted with the problem must never swallow a product containing peanut even in the minutest amount; this calls for great vigilance not only on the part of the child, but particularly the parent. Thankfully, those with the allergy are not affected by peanut vapor (smell). Other foods can cause similar allergic reactions similar to peanuts, and usually for the same reason: creation of antibodies, mast cells, and histamine. Examples are other varieties of nuts, seafood, wheat (gluten in wheat), and soy. Food intolerances are sometimes the result of a missing enzyme; for example, intolerance for dairy products may be due to the absence of lactase, the enzyme that breaks down lactose, also called milk sugar. Many people suffer allergies from the pollen of weeds, plants, flowers, etc.; again because of the body's production of histamine. Ragweed, a tall leafy weed common to the eastern and midwestern states, is notorious for causing miserable sinus infections in 10–20% of Americans. No surprise that ragweed is so obnoxious since each plant produces one billion pollen spores. The allergy to bee stings should not be overlooked. Most people are allergic to a sting from one of the many different types of bees and wasps. The typical reaction is swelling and itching at the site of the sting, but occasionally a person is extremely allergic and suffers anaphylactic shock. Finally, there is the allergic reaction to the plants poison ivy, sumac, and oak. Poison ivy and sumac are common in the eastern and midwestern states while poison oak occurs in the West. In the case of these plants the symptoms are not caused by histamine, but from a liquid secretion called urushiol, which rubs on the skin when touched. Contact with urushiol can lead to a type of dermatitis, with mild to severe rashes, extreme itching, and blistering, occasionally leading to serious medical problems. This is particularly true if smoke from burning plants is breathed into the lungs. In such cases, the pain is severe and there is the possibility of fatal respiratory problems.

Pre-Laboratory Questions | 32

1. List the biochemical constituents of the peanut and the test you will use to detect each of them.

2. Draw the structure of a typical peanut oil molecule, using the most probable fatty acid residues.

3. Do the monomer units (xylose) in xylan have the alpha or beta configuration? Is xylose a reducing sugar? Is it a pentose or a hexose?

4. List the functional groups that are present in acrolein. What physical property of acrolein makes it easily detected?

5. Define the term *pentose*. Distinguish between an aldopentose and a ketopentose. Distinguish between an aldopentose and an aldohexose.

EXPERIMENT

1. The Kernel

A. Isolation of Peanut Oil

Separate the shells and kernels of two large roasted peanuts. Save the shells for a later part of the experiment. *Weigh* the four kernels, without the skins, Record the weight in line 1A, 1. Grind them with a pestle in a mortar or a dry evaporating dish. Note the consistency of the ground peanuts. (Answer Question 1A, 2–3 on the Observation and Results sheet.) Now add 10 mL of methylene chloride, CH_2Cl_2, to the mortar,[1] and continue grinding for a short time. *Cover the mortar with a watch glass to prevent evaporation while you are obtaining filter equipment.*

Set up a filter (figure 32.1) by placing fluted[2] filter paper in a funnel whose stem extends into a dry, weighed (tared/zeroed) 50-mL Erlenmeyer flask. Transfer the extraction solution from the mortar to the funnel.[3] *Cover the funnel with a watch glass.* After a few minutes, press lightly on the top rim of the filter paper in order to increase the rate of filtration. Proceed with the analysis of the peanut shell (part 2) while the filtration continues. When the rate of dripping has slowed considerably, remove the filter paper and carry it *to the hood.* Use your spatula to transfer the peanut meal to a piece of paper towel. Spread out the meal and allow it to remain under the hood until it dries.

Dispose of the wet filter paper in the appropriate container under the hood.

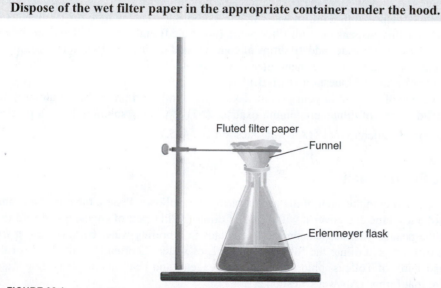

Fluted filter paper

Funnel

Erlenmeyer flask

▲ **FIGURE 32.1** Fluted filter apparatus.

---

[1] Work under the hood when using methylene chloride.

[2] Obtain the fluted filter paper from the stockroom or prepare one according to your instructor's directions.

[3] Perform the transfer quickly so the meal does not run down the side of the mortar and contaminate the filtered solution.

Place the 50-mL Erlenmeyer flask on a hot plate *under the hood* and evaporate the methylene chloride. When the boiling ceases, cool and weigh the flask and peanut oil.

Weigh the oil and record in 1A on the Observation and Results sheet. Calculate the percentage of oil and record in 1A.

B. Analysis of the Peanut Oil

Now perform some tests to establish that the liquid that you isolated is indeed a vegetable oil. For example, to see whether unsaturated fatty acid residues are present in the triglyceride add 1 drop of 2% potassium permanganate, $KMnO_4$, to 2 drops of peanut oil in 2 mL of ethanol, and observe whether the purple color changes to brown in a short time. (See Experiment 19 for a discussion of the reaction of $KMnO_4$ with alkenes.) (Answer Question 1B, 1–3.)

Prove that your peanut oil is a triglyceride by employing the acrolein test: Place a few drops of the peanut oil and an equivalent amount of solid potassium hydrogen sulfate, $KHSO_4$, in a dry test tube. Heat the mixture *strongly* in the flame. Observe the vapors, and waft them toward your nose. Do you detect the odor of acrolein? (Answer Question 1B, 4.)

C. Analysis of the Peanut Meal

Use Fehling's test to detect the presence of carbohydrates in the peanut meal. With a stirring rod, grind a portion of the meal the size of a large pea with 2 mL of hot water in a small beaker. Then dilute to 10 mL with water and divide the resulting suspension into two parts in test tubes. On one portion run Fehling's test *at once*. On the other portion of the suspension, attempt hydrolysis before doing Fehling's test: Add 1 mL of concentrated HCl to the suspension and heat it in a boiling water bath for 10 minutes. Neutralize the solution with 10% NaOH (litmus test), and then perform Fehling's test. Report your observations.

Use the xanthoproteic test to prove that the peanut contains protein. With a stirring rod, grind a portion of meal the size of a large pea with 5 mL of water in a test tube, and heat the tube for 10 minutes in a boiling water bath. Filter this suspension with filter paper (not fluted) and a funnel. To 1 mL of the filtrate (save the remainder of the filtrate) add 10 drops of concentrated nitric acid. Heat this carefully in the flame for a few minutes. Cool the filtrate, neutralize it with 10% NaOH solution, and note its color. Do peanuts contain protein? (Answer Question 1C, 1–3.)

Now test for the presence of calcium in your peanut. To another 1-mL portion of the filtrate used for the xanthoproteic test, add 1 mL of dilute ammonium oxalate, $(NH_4)_2C_2O_4$, solution. Does a precipitate form? Do peanuts contain calcium ions? (Answer Question 1C, 4–5.)

2. Analysis of the Peanut Shell

To test for the presence of xylan in the shell of the peanut proceed as follows. Press a quarter of a peanut shell into the bottom of a test tube and cover it with 3 mL of dilute HCl (1 part of concentrated acid to 3 parts of water). Heat the peanut shell suspension for 15 minutes in a boiling water bath, stirring it frequently. Filter the suspension, and dilute the filtrate with 3 mL of water. Thoroughly mix 1 mL of the hydrolysis solution and 1 mL of Tollens' phloroglucinol test reagent, and heat this in the boiling water bath. Observe the color that forms. (Answer Question 2, a–f.)

> **Dispose of all waste from part B and C and part 2 of the experiment in the appropriate containers.**

QUESTIONS AND PROBLEMS

1. Would survival be possible on a diet composed solely of peanuts and water? Explain.

2. Since the peanut shell contains two carbohydrates, why are peanut shells not ground up into flour and used as a food source?

3. Compare the composition of peanut oil, corn oil, and soybean oil with respect to (a) types and percentages of fatty acid residues, (b) the similarity of the "alcohol portion," and (c) uses.

4. Discuss the chemical reactions that occur in the process of digesting peanuts. Use equations to describe the reactions. (Hint: Recall the biochemical constituents of the peanut and what happens to these constituents during digestion.)

5. Most commercial peanut butter contains an emulsifying agent. What is the purpose of this compound?

6. Peanut butter becomes rancid when it is stored for some time. Which constituent is responsible for the development of rancidity? Describe the reaction that is responsible for rancidity.

EXPERIMENT

Analysis of an Important Food: The Peanut

32

OBSERVATIONS AND RESULTS

1. The Kernel

   A. Isolation of Peanut Oil

      1. Weight of materials and calculations   _____ g

        Weight of the four peanut kernels   _____ g

        Weight of the peanut oil   _____ g

        Percentage of oil in the kernels   _____ %

      2. Describe the consistency of ground peanuts.

      3. What is the commercial name for ground peanuts?

      4. Does the percentage of your isolated oil agree with the value reported in the discussion section? If not, why not?

5. Describe the color, odor, and viscosity of the peanut oil.

B. *Analysis of the Peanut Oil*

1. Did $KMnO_4$ react with your peanut oil? What is your evidence for a reaction?

2. What does the $KMnO_4$ test suggest about possible fatty acid residues in the triglyceride molecules of peanut oil?

3. Write an equation for the reaction of a typical peanut oil molecule with $KMnO_4$.

4. Did the acrolein test confirm that your liquid was a triglyceride? Explain.

C. *Analysis of the Peanut Meal*

1. Report your observations on Fehling's test.

Before hydrolysis:

After hydrolysis:

2. Is the carbohydrate in peanut meal a reducing or a nonreducing sugar? What type of carbohydrate (monosaccharide, disaccharide, or polysaccharide) is suggested by this test? What is your evidence?

3. Draw a structure of a typical molecule of this carbohydrate.

4. Did the xanthoproteic test establish that peanuts contain protein? Explain.

5. What amino acid residues in peanut protein are confirmed by this test?

6. What did you observe when ammonium oxalate was added to the aqueous extract (filtrate) of peanut meal? Write an ionic equation for the reaction that occurred.

7. Give two reasons why it is necessary to have calcium in your diet.

2. Analysis of the Peanut Shell

a. What color did you observe when the solution from the hydrolysis of the peanut shell was heated with phloroglucinol test reagent? Did the test confirm the presence of xylose?

b. Draw the structure of xylose.

c. Does this test also suggest that xylan is present in the peanut shell? What is your evidence?

d. Draw a segment of the structure of a molecule of xylan. Circle the monomer units, and indicate (using arrows) the acetal bonds.

e. Some glucose was formed during the hydrolysis of the peanut shell. What was the source of the glucose?

f. Did the glucose give a positive Tollens' phloroglucinol test? Would Fehling's test confirm unequivocally the presence of this carbohydrate? Explain.

Milk and Its Principal Components | 33

OBJECTIVES

1. To become familiar with the principal chemical components of milk.
2. To apply reactions that were studied in previous experiments to an analysis of milk.
3. To isolate two components of milk, casein and butterfat.

DISCUSSION

Milk or milk products (butter and cheese) from cows, goats, and sheep have been a major food source from earliest recorded history. Across time milk production spread from its apparent source in the grasslands of southwestern Asia to other parts of the world. In earlier times, each family had sufficient goats and sheep to supply milk to meet its own needs. With the establishment of cities, cows became the principal source of milk because of the greater volume of production. Milk from cows, goats, or sheep is chemically quite similar; sheep's milk is somewhat richer in protein and fat, and cow's milk has the highest milk sugar (lactose) content.

Some of the major components of milk are water (87%), fat (3.9%), protein (3.5%), and lactose (4.9%). The following vitamins are present in milk in small but important amounts (mg/kg): thiamine (B_1), 0.44; riboflavin (B_2), 1.75; nicotinic acid, 0.94; pyridoxine (B_6), 0.64; pantothenic acid, 3.46; biotin, 0.31; cyanocobalamin (B_2), 0.0043. Carotene (vitamin A) is present at 156 IU/100 g. Milk contains vitamin C (1 mg/kg), but this vitamin is easily destroyed by heating during the pasteurization process. Nowadays Vitamin D is normally added to milk. The concentration of calcium is high, 118 mg/100 g. Many proteins and enzymes are found in milk. A prominent protein is casein, which exists as calcium caseinate in a fine dispersion and is responsible for the white color of milk. The carotenes give the yellow color to butterfat. The greenish yellow color of whey, the watery part after milk sours, is produced by the presence of riboflavin.

In today's experiment you will isolate two of the principal components of milk, casein and butterfat, and conduct tests for some of the others: lactose, some proteins, calcium, and phosphate. You will also compare fresh and sour milk. The procedures involved in the experiment are outlined as follows:

Because of the length of the experiment, your instructor may recommend that you work together in pairs.

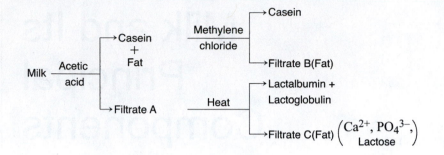

Pre-Laboratory Questions | 33

1. Write structures and names for some of the monomers in casein.

2. Write a brief paragraph describing how you will separate casein and butterfat.

3. Explain why sour milk has a sour taste.

4. Obtain the structure of lactose from your textbook. Circle and name the monosaccharides in this disaccharide. Is lactose a reducing sugar? Explain.

5. Describe the acrolein test and represent it with a chemical equation.

EXPERIMENT

1. Precipitation of Casein and Butterfat

Add 3 mL of 10% acetic acid to 25 mL of whole milk in a 100-mL beaker. Stir the mixture while heating to 50°C; continue heating and stirring on a ring stand until curds of casein appear (about 5 minutes at 50°C with stirring). Filter (15 cm) through a fluted filter (see Experiment 33) in a funnel, collecting the filtrate (A) in a 250-mL beaker. Wash the precipitate (casein-butterfat) in the filter with a 10-mL portion of water, combining the filtrate with (A). Save filtrate (A) for part 3 and the precipitate for part 2. (Answer Question 1, a–b on the Observation and Results sheet.)

2. Separation of the Casein and Butterfat

Prepare a steam bath as shown in figure 33.1 with a 400- to 600-mL beaker and an appropriately sized watch glass. Fill the beaker half full of water. Transfer the precipitate of casein-butterfat from part 1 from the filter paper to the watch glass and distribute the casein around the watch glass. Place the watch glass on top of the beaker and boil the water vigorously. In a few minutes water will separate from the solid. Remove as much of the water as possible with a dropping pipet. Continue to heat the contents of the watch glass until the solid appears dry. Meanwhile proceed to part 3.

Transfer the dry precipitate of casein-butterfat to a test tube (*1.5 × 15 cm*). Add 10 mL of methylene chloride to the test tube under the hood. Vigorously stir the contents by working the solid against the sides of the test tube with your stirring rod. Butterfat is trapped in the casein and will be extracted only as fresh surface is exposed to the solvent.

Prepare a fluted filter and weigh (tare/zero) a clean, large (*2.5 × 20 cm*), dry test tube. Filter the above mixture through a clean fluted filter in a funnel and catch the filtrate (B) in the test tube. Wash the precipitate of casein once with 5 mL of methylene chloride, and collect the filtrate in the same test tube with filtrate B. Filtrate B contains butterfat dissolved in methylene chloride. Spread the fluted filter with the casein to dry under the hood. When all of the methylene chloride has evaporated, weigh the casein and record the weight. (Answer Question 2a and perform calculation.) Perform the xanthoproteic test

▲ **FIGURE 33.1** Steam bath for drying casein.

(Experiment 30) for protein on a small portion (size of a large pea) of the casein: Suspend the casein in 1 mL of H_2O and add 10 drops of nitric acid. Heat for several minutes in a boiling water bath. (Answer Question 2b.)

Place the test tube with the methylene chloride solution (filtrate B) in a boiling water bath under the hood. (Large beakers on hot plates will be provided.) Frequently shake the test tube in the water bath until all of the solvent has boiled off. You should not be able to smell any more methylene chloride. Cool the test tube and contents in an ice bath. When the contents have become viscous or solidified, dry the outside of the test tube, zero the balance, and weigh the butterfat. (Record the weight of the butterfat in 2c and answer 2, d–f.)

Again cool the test tube in the ice bath and remove some of the solid with a spatula. Note the appearance and odor. To establish that the solid is a glyceride, perform the acrolein test as described in Experiment 32. (Answer Question 2, g–h.)

3. Isolation of Lactalbumin and Lactoglobulin

Gently boil the filtrate (A) from the precipitation of casein and butterfat in a large test tube (2.5×15 cm) until you observe the formation of a suspended precipitate (about 10 minutes). This precipitate is the result of the coagulation of the proteins lactalbumin and lactoglobulin. Isolate these proteins by filtration with a fluted filter (9 cm). Expect a precipitate the size of a small pea. Save the filtrate (C) for later tests. Remove the lactalbumin and lactoglobulin from the tip of your filter paper with your spatula. (Answer Question 3a.)

Establish that proteins are present using the biuret test (Experiment 30): Transfer some of the solid protein to 2 mL of water in a test tube and stir thoroughly to dissolve as much as possible. Now perform the biuret test. (Answer Question 3b.)

4. Test for Calcium (Ca^{2+}) in Milk

To 1 mL of filtrate C from part 3, add 1 mL of dilute ammonium oxalate, $(NH_4)_2C_2O_4$. Did a precipitate form? Consult Experiment 32 for information on the ammonium oxalate test for calcium. (Answer Question 4, a–c.)

5. Test for Phosphate, PO_4^{3-}, in Milk

The presence of phosphate, PO_4^{3-}, can be detected by reacting ammonium molybdate, $(NH_4)_2MoO_4$, with phosphate and observing the formation of a yellow precipitate of ammonium phosphomolybdate, $(NH_4)_3PO_4 \cdot 12 MoO_4$. Perform this test for phosphate, as described in Experiment 16, on 3 mL of filtrate C from part 3. (Answer Question 5, a–c.)

6. Detection of Lactose in Milk

Perform the Fehling's test for carbohydrate (see Experiment 26) on 5 drops of filtrate C. (Answer Question 6, a–c.)

Dispose of all waste from each part of this experiment in the appropriate containers.

7. Comparison of Fresh and Sour Milk

Determine the pH of fresh milk and sour milk. Use wide-range pH paper to locate the approximate pH and narrow-range paper to determine the exact pH. (Record your results and answer Question 7, a–c.)

QUESTIONS AND PROBLEMS

1. Write equations showing how microorganisms convert lactose to lactic acid. Use your textbook to review the glycolysis sequence. Assume that microorganisms can convert galactose to glucose.

2. Describe how you could prove that sheep's milk contains more protein than cow's milk.

3. Suggest a procedure that a dairy company might use for making nonfat milk from whole milk.

4. Compare the protein content of fresh and sour milk.

5. Discuss how dried milk might be made from fresh milk. Would you expect reconstituted dried milk to be as nutritional as fresh milk? Explain.

6. Defend the following statement from a chemical standpoint: "Milk is an excellent food source for development of bones and teeth."

EXPERIMENT

Milk and Its Principal Components

33

OBSERVATIONS AND RESULTS

1. Precipitation of Casein and Butterfat

a. Describe the appearance of the casein.

b. The butterfat precipitated with the casein. Where is the butterfat at this point in the experiment?

2. Separation of the Casein and Butterfat

a. Weight of casein: _____ g. Based on a density of 1.0 g/mL for milk, calculate the percentage of casein in milk.

b. Did the xanthoprotic test confirm that casein is a protein? _____ Explain.

    c. Weight of butterfat: _____ g

    d. Based on a density of 1.0 g/mL for milk, calculate the percentage of butterfat in milk.

    e. Does your value agree with the butterfat content on the label? _____ Explain.

    f. Describe the odor and appearance of your butterfat.

    g. Did the acrolein test establish that butterfat is a triglyceride? What was the evidence?

    h. Draw the structure of a molecule that would be typical of one found in butterfat.

3. Isolation of lactalbumin and lactoglobulin

    a. Describe the appearance of the lactalbumin and lactoglobulin.

    b. Did the biuret test confirm that you had isolated proteins? _____ Explain.

4. Test for Calcium, Ca^{2+}, in Milk

a. What happened when you added ammonium oxalate, $(NH_4)_2C_2O_4$, to filtrate C?

b. Write an equation for the reaction that took place.

c. Does this result prove that milk contains calcium, Ca^{2+} ? _____ Explain.

5. Test for Phosphate, PO_4^{3-}, in Milk

a. Did the ammonium molybdate, $(NH_4)_2MoO_4$, establish the presence of phosphate, PO_4^{3-}, in milk? _____ Explain.

b. Write an equation for the reaction.

c. Milk is an important source of calcium and phosphate in the diet. What is a principal use for these ionic materials (Ca^{2+} and PO_4^{3-}) in the body?

6. Detection of Lactose in Milk

a. Did the Fehling's test detect lactose in milk? _____ Explain.

b. Is lactose a reducing sugar? _____ Explain.

c. Write an equation for the reaction between lactose and Fehling's solution.

7. Comparison of Fresh and Sour Milk

a. pH of fresh milk: _____ pH of sour milk: _____ Which type of milk has

the lower pH _____ ? The greater acidity _____ ? The greater H^+

concentration _____ ? What organic (carboxylic) acid produces the H^+ in sour milk_____ ?

What is its source?

b. Are there "curds" in the sour milk? _____ What caused the milk to coagulate
(produce "curds")?

c. What are the principal compounds in the "curds"? _____

Optical Activity: A Study of Chiral Molecules | 34

OBJECTIVES

1. To develop an understanding of chirality by building models of chiral molecules and drawing three-dimensional structures.

2. To become familiar with common materials that contain chiral components.

3. To observe the practical implications of chirality and optical activity by using the polarimeter.

4. To consider the importance of chirality in biochemical reactions.

DISCUSSION

Molecules may be chiral,[1] or handed, just like your right and left hands. Chiral molecules have the same relationship to each other as your hands do. If you hold your right hand up to a mirror, the reflection looks like your left hand. The same is true for a pair of chiral molecules, as shown for the 2-chlorobutanes.

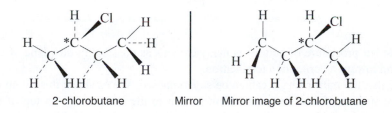

2-chlorobutane   Mirror   Mirror image of 2-chlorobutane

Both the 2-chlorobutane on the left and its mirror image on the right exist as distinct molecules. They are isomers—actually stereoisomers (stero meaning in space). These particular types of stereoisomers, where the sole structural difference is that of mirror images, are called enantiomers. They are also called optical isomers because the only experimental difference they show is that they rotate plane-polarized light in opposite directions. Therefore, we say that a chiral molecule (one of a pair of enantiomers) is optically active. We will return to the subject of plane-polarized light later in this discussion.

Now let us take a more detailed look at a chiral molecule. A chiral molecule normally contains a chiral carbon. A chiral carbon, like the asterisked C in the 2-chlorobutane enantiomers, is a carbon with four different atoms or groups around it. In the case of 2-chlorobutane these groups are chlorine, hydrogen, methyl, and ethyl. In drawing chiral molecules, we place as many carbon atoms as possible in the

---

[1] *Chiral* is derived from the Greek word *cheir,* meaning hand.

Copyright © 2013 Pearson Education, Inc.   419

plane of the paper and connect the other atoms with dashed lines (----) if they are below the plane of the paper and with wedges (◄) if they are above the plane of the paper. All bonds from saturated carbons have 109° between them and are directed in space toward the corners of a regular tetrahedron, as shown in the structure of methane (figure 34.1). We have attempted to show the correct bond angles and the tetrahedral arrangements around the carbons in our three-dimensional drawings of the 2-chlorobutanes using dashed and wedged lines.

What are the tests or observations that you can make on the structure of a molecule to determine whether it is chiral? The first test is to determine whether it has a chiral carbon; that is, does it have a carbon with four different groups around it? The second is to draw the three-dimensional, mirror-image structure of the molecule and then, in your mind, superimpose it on (place it over the top of) the original molecule. If the mirror image cannot be superimposed so that it is identical, the molecule is chiral and will show optical activity. Consider, for example, aspartic acid, a very important amino acid, whose two-dimensional structure is

$$HOOC-CH_2-\overset{*}{C}H-COOH$$
$$|$$
$$NH_2$$

Does aspartic acid show chirality (is it chiral) and will it show optical activity? It passes the first test of having a chiral carbon (indicated by *), with four different groups attached: amino ($-NH_2$), hydrogen, $-CH_2COOH$, and carboxyl ($-COOH$). To apply the second test, we will draw the structure of aspartic acid in three dimensions and determine if the mirror image is superimposable.

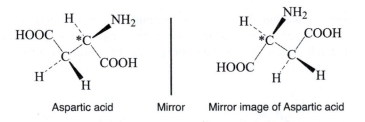

Aspartic acid   Mirror   Mirror image of Aspartic acid

Again we follow our procedure of putting as many carbons as possible in the plane of the paper and directing each bond toward a corner of a tetrahedron.

We find here that the mirror image cannot be superimposed. We establish this by, in our mind, lifting the mirror-image structure from the paper and moving it to the left directly on top of the aspartic acid molecule. When you do this, you will note that the amino group and hydrogen superimpose, but the CH_2COOH is over the COOH and the COOH is over the CH_2COOH. The superimposability test always requires one other operation: lift the mirror image from the plane of the paper as you just did, but now turn it over 180°. Now you will observe that the CH_2COOH groups and the COOH groups superimpose, but the amino groups and hydrogens do not. Therefore, since the mirror image is nonsuperimposable, the compound is chiral and will show optical activity. There is one situation in which chiral molecules are

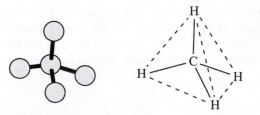

▲ **FIGURE 34.1** Tetrahedral structure of methane.

present but do not rotate plane-polarized light. This is the case of a 50:50 mixture of enantiomers or mirror-image molecules. Each molecule is chiral and actually rotates plane-polarized light, but its rotation is cancelled by its mirror image. Such a mixture is called a racemic mixture.

Now let us return to an examination of the experimental difference between a pair of chiral molecules or enantiomers. As we mentioned earlier, the only difference between enantiomers is the fact that they rotate plane-polarized light equally, but in opposite directions. All other physical properties and chemical reactions are identical. Now let us define exactly what plane-polarized light is. Plane-polarized light is light that is vibrating in one direction only. The instrument that is used to generate the plane-polarized light and to determine the magnitude of the rotation of the plane-polarized light is called a polarimeter. Your instructor will demonstrate the use of the polarimeter in today's experiment and explain its operation. The magnitude of the rotation, called the specific rotation and represented by $[\alpha]$, is given in degrees plus (+) or minus (−), depending on whether the light is rotated clockwise (+) or counterclockwise (−). In the case of our first example, the 2-chlorobutanes, $[\alpha]$ has been determined experimentally to be +50° and −50°. Similarly, the enantiomers of the aspartic acids, in our second example, show rotations of +24° and −24°. A plus (+) enantiomer is also known as dextrorotatory (d−), and a minus (−) enantiomer is known as levorotatory (l−).

The field of optical activity or chirality is not simply an area of intellectual curiosity or of interest only to theoretical chemists. It is of great significance in the study of the molecules of living organisms because most of these molecules are chiral. This includes such molecules as carbohydrates, proteins, nucleic acids, steroids, and enzymes. Chirality is one of nature's most important tools for recognition and distinction. Chirality is the basis whereby enzymes recognize their substrates. Chirality ultimately controls reaction pathways. Therefore, you can see why a working knowledge of chirality is of such great importance to scientists in the fields of biochemistry and molecular biology, as well as to organic chemists.

In today's experiment, you will examine the physical and chemical properties of several optical isomers (enantiomers). You will build models to represent them and will learn to draw their structures in three dimensions. You will learn about the physical and chemical properties of optical isomers. You will see how the polarimeter operates, and how the magnitude of plane-polarized rotation is determined.

Pre-Laboratory Questions | 34

1. Define the following: chiral molecule, plane-polarized light, superimposable, optical activity, racemic mixture, and polarimeter.

2. Discuss the relationship between the following terms: enantiomer and optical isomer; chiral molecule and enantiomer.

3. Name the sources in nature for (+)-, (−)-, and (±)-carvone.

4. Identify the functional groups in the carvone molecule.

5. Draw the three-dimensional structure of a chiral molecule that is not discussed in this experiment.

EXPERIMENT

1. Examination of Two Simple Chiral Molecules: (+)- and (−)-Alanine

The two-dimensional structure for alanine is

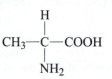

Copy this structure of alanine on your answer sheet and put an asterisk by the chiral carbon. (Answer question 1a on the Observation and Results sheet.)

Now, using the model of a chiral carbon provided by your instructor, construct models of (+)- and (−)-alanine using copper wire and different colored corks. Cut two 6″ pieces of copper wire. Twist them together several times at their centers until they give a firm contact and do not slip. Bend the wires so that each end is directed toward a corner of a tetrahedron as shown in your instructor's model. Cut each wire to approximately the same length. Punch 6-mm holes in four corks with an opened paper clip and attach the corks to your wire model as shown in figure 34.2. Using a felt pen, color a portion of each cork with a different color in order to represent the four groups in alanine: CH_3, $COOH$, H, and NH_2. Now imagine what the mirror image of your model alanine would look like and construct that model too. You now have models for (+)- and (−)-alanines. Are they superimposable? (Answer Question 1, b–c.)

Have your instructor approve your models before proceeding. Approval by your instructor means that your models are constructed correctly, you understand the concept of superimposability, and you are able to draw the two enantiomers correctly in three dimensions. Compare the chemistry of (+)-, (±)-, and (−)-alanines by determining the pH of a solution of each with pH paper. (Record the pH for each solution and answer Question 1e on the Observation and Results sheet.)

Dispose of all waste in the appropriate container.

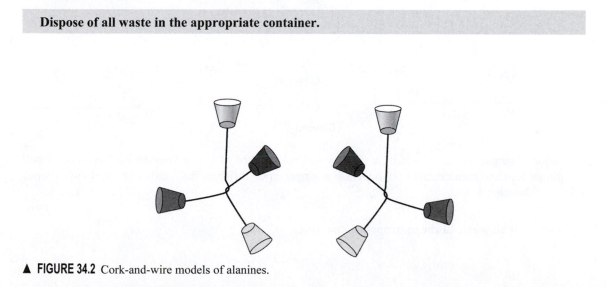

▲ **FIGURE 34.2** Cork-and-wire models of alanines.

2. Physical Properties of the Optically Isomeric Menthols

The two-dimensional structure of menthol is

$$CH_3-CH \begin{matrix} CH_2-CH_2 \\ \diagup \qquad\quad \diagdown \\ \qquad\qquad CH-CH(CH_3)_2 \\ \diagdown \qquad\quad \diagup \\ CH_2-CHOH \end{matrix}$$

You may remember that you studied this camphor-smelling compound in Experiment 20. Draw the two-dimensional structure of menthol on your answer sheet and identify the three chiral carbons with asterisks. You will draw some of the enantiomers of menthol when you answer Question 1 in the Questions and Problems section. (Answer Question 2a.)

Observe the odors of the (+)-, (−)-, and (±)-menthols. Are they identical? Take the melting points of the menthols as described in Experiment 18. Use a 250-mL beaker, half filled with water. Record your melting points. Remember that melting points of the same compound may have a range of 3°C because of variations in heating rates. Change water after each melting point so you will not have to wait for it to cool down. (Answer Question 2, b–d.)

3. Optical Isomers in Some Common Materials

In this part of the experiment, you will compare the odors of three common materials and relate these odors to the authentic (pure) chemicals that are responsible for these odors. Proceed by comparing the odors of crushed spearmint leaves and crushed caraway seeds (or dill). Can you tell a difference? List other materials in which you have smelled these odors.

The compounds responsible for the odors of caraway (or dill) and spearmint are (+)- and (−)-carvones, respectively. Smell the authentic (−)- and (+)-carvones on the sideshelf. Do they smell like spearmint and caraway? The two-dimensional structure of carvone is shown below. On your answer sheet, draw the two optical isomeric carvones in three dimensions, putting the cyclohexene ring (and methyl group and oxygen) in the plane of the paper and the groups at the chiral carbon above and below the plane of the paper.

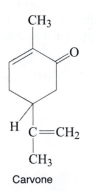

Carvone

The odor of ginger root is caused by a racemic (50:50) mixture of (+)-carvone and (−)-carvone. Smell the ginger root and then describe the odor. Does ginger smell different from either (+)- or (−)-Carvone? (Answer Question 3, a–d.)

Dispose of all waste in the appropriate container.

4. Rotation of Plane-Polarized Light by Sucrose—A Demonstration of the Polarimeter

In this part of the experiment, your instructor will demonstrate the polarimeter by determining the specific rotation $[\alpha]$ of a 5% sucrose solution. Sucrose (table sugar) is an optically active disaccharide (see Experiment 26). The specific rotation $[\alpha]$ is related to the observed rotation (from the polarimeter) as follows.

$$[\alpha] = \frac{\text{Observed rotation (polarimeter)}}{\text{Length of tube (decimeters)} \times \text{concentration (g/mL)}}$$

(As the demonstration proceeds, record the data and answer Question 4.)

QUESTIONS AND PROBLEMS

1. Menthol, which you studied in this experiment, has three chiral carbons. Draw the three-dimensional structures of three different pairs of enantiomers of menthol by changing the directions in space of the groups at the chiral carbons. (Four pairs of enantiomers actually exist.) In all cases, draw the cyclohexane ring in the plane of the paper as directed in the experiment.

2. Discuss how it is possible for enzyme molecules, which are chiral, to recognize and react with only one of the enantiomers in a (+)-, (−)- pair.

3. The two-dimensional structure of tartaric acid is shown below. Draw the three-dimensional structures of the different stereoisomers for tartaric acid and describe their relationships.

$$\begin{array}{c} \quad\; H \;\; H \\ \quad\; | \;\;\; | \\ HOOC-C-C-COOH \\ \quad\; | \;\;\; | \\ \quad\; OH \; OH \end{array}$$

EXPERIMENT

Optical Activity: A Study of Chiral Molecules

34

OBSERVATIONS AND RESULTS

1. Examination of Two Simple Chiral Molecules: (+)- and (−)-Alanines

a. Draw the two-dimensional structure of alanine and identify the chiral carbon with an asterisk.

b. Is the model of the mirror image of your alanine superimposable on the original alanine?
_____ . Are your two models identical? _____ Explain.

c. Draw three-dimensional structures of *your two models* of the optically active alanines. Show each cork and indicate its color.

d. Your instructor's approval of your models: _____
(Signature)

e. Record the pH of the solutions of the alanines: (+)- _____ ; (±)- _____ ;
(−)- _____ . Are the acid-base properties of enantiomers the same? _____
How do the acid-base properties of the enantiomers compare to those of the racemic mixture?

2. Physical Properties of the Isomeric Menthols

a. Draw a two-dimensional structure of menthol below and identify each chiral carbon with an asterisk.

b. Do the enantiomeric menthols have the same odor? _____ If not, describe their odors.

c. Melting points of the menthols: (+)- _____ °C; (±)- _____ °C; (−)- _____ °C.
   Which of the three melted within a ± 3°C range of the reported value for (+)- or (−)-menthol?

d. If any of them is outside the range, suggest a possible explanation.

3. Physical Properties of the Optically Isomeric Carvones

a. Could you smell a difference between the odors of spearmint and caraway? _____ .
   It is reported that about one out of ten people cannot detect a difference. In what materials have you smelled the odors of spearmint and caraway?

b. Did the authentic (−)- and (+)-carvones smell like caraway and spearmint? _____
   Draw the three-dimensional structures of (+)- and (−)-carvones.

c. Describe the odor of ginger root. Did it smell different from either (+)- or (−)-carvone?

d. In your own words, suggest an explanation at *the molecular level* for the difference in the odors of spearmint, caraway, and ginger.

4. Rotation of Plane-Polarized Light by Sucrose—A Demonstration of the Polarimeter

The observed rotation for sucrose (value obtained from the polarimeter): _____ °.

The specific rotation [α] for sucrose _____ °. The reported (literature) value for

the specific rotation [α] for sucrose: _____

Appendix

TABLE A.1 Activity (or Replacement) Series of Some Metals

Metals

Most active metals; prepared by electrolysis	K Ba Ca Na Mg Al	Liberate H_2 from water
Oxides reduced by aluminum or carbon	Mn Zn Cr Fe	Liberate H_2 from acids
Oxides reduced by H_2 or CO	Cd Ni Sn Pb H	React slowly with acids
Oxides decomposed by heat alone; least active metals	Cu Hg Ag Au Pt	React only with strong oxidizing acids; no H_2 produced

TABLE A.2 Concentrations of Common Acid and Base Solutions

Reagent	Formula	Molarity (M)	Density (g/mL)	Percent Solute (%)
Acetic acid (glacial)	$HC_2H_3O_2$	17	1.05	99.5
Acetic acid (dilute)		6	1.04	34
Hydrochloric acid (concentrated)	HCl	12	1.18	36
Hydrochloric acid (dilute)		6	1.10	20
Nitric acid (concentrated)	HNO_3	16	1.42	72
Nitric acid (dilute)		6	1.19	32
Sulfuric acid (concentrated)	H_2SO_4	18	1.84	96
Sulfuric acid (dilute)		3	1.18	25
Ammonium hydroxide (concentrated)	NH_4OH	15	0.90	58
Ammonium hydroxide (dilute)		6	0.96	23
Sodium hydroxide (dilute)		6	1.22	20

TABLE A.3 Relative Strengths of Some Common Acids and Bases (in order of decreasing strength)

Acids	Formula
Perchloric acid	$HClO_4$
Hydrochloric acid	HCl
Nitric acid	HNO_3
Sulfuric acid	H_2SO_4
Oxalic acid	$H_2C_2O_4$
Hydrogen sulfate ion	HSO_4^-
Sulfurous acid	H_2SO_3
Phosphoric acid	H_3PO_4
Hydrofluoric acid	HF
Ferric ion	Fe^{3+}
Nitrous acid	HNO_2
Acetic acid	$HC_2H_3O_2$
Aluminum ion	Al^{3+}
Carbonic acid	H_2CO_3
Boric acid	H_3BO_3
Bases	
Sodium hydroxide	$NaOH$
Calcium hydroxide	$Ca(OH)_2$
Ammonia	NH_3

TABLE A.4 Color Changes, with pH Intervals, of Some Important Indicators

Indicator	pH Interval	Color Change
Methyl violet	0.2–3.0	Yellow to violet
Thymol blue	1.2–2.8	Red to yellow
Methyl orange	3.1–4.4	Red to orange to yellow
Bromphenol blue	3.0–4.6	Yellow to blue-violet
Congo red	3.0–5.0	Blue to red
Bromcresol green	3.8–5.4	Yellow to blue
Methyl red	4.4–6.2	Red to yellow
Bromcresol purple	5.2–6.8	Yellow to purple
Litmus	4.5–8.3	Red to blue
Bromthymol blue	6.0–7.6	Yellow to blue
Phenol red	6.8–8.2	Yellow to red
Thymol blue	8.0–9.6	Yellow to blue
Phenolphthalein	8.3–10.0	Colorless to red
Thymolphthalein	9.3–10.5	Yellow to blue
Alizarin yellow R	10.0–12.0	Yellow to red
Indigo carmine	11.4–13.0	Blue to yellow
Trinitrobenzene	12.0–14.0	Colorless to orange